W9-BGQ-202

Wholeness *and* Disintegration *in*
Ecology *and* Science

ACKNOWLEDGMENTS

The author would like to thank the following individuals who all contributed, in their various ways, to the production of this book: Colin Burrows, Guenter Getzinger, Silvia Lozeva, Brian Martin, David Mercer, Harald Rohracher, John Schuster, John Schumacher, Bernhard Wieser.

The research and writing for this book was carried out at the Department of Science, Technology and Society at the University of Wollongong, Wollongong, Australia, and at the Institute for Advanced Studies in Science, Technology and Society, Graz, Austria.

Portions of Chapters Two and Eight of this book have been adapted from an article previously published in *Studies in the History and Philosophy of Science*, Part C, Vol 29 (1998), No 1, pp137-164 (with permission from Elsevier Science).

CONTENTS

INTRODUCTION

What is the nature of Nature? Is it harmonious and graceful? Is it divine and providential? Is it cruel and wild? These days, answers to the question 'What is the nature of Nature?' often reverberate with claims of natural unity: every living thing in nature, and this includes humanity, is united.

The unity of nature idea appears in many different forms; in the speeches of environmentally conscious politicians, in the academic writings of scientists, in the sermons of theological leaders, and in the slogans of social activists. The most common site, however, for the description of the unity of nature is ecology and environmentalism. Ecology and environmentalism show us how the Earth's biological members are involved in intricately entangled relationships with one another. Ecology and environmentalism also tell us how all the world's material things flow through ecological systems in complex cycles, uniting these systems as they flow.

Much of the time, the unity of nature is expressed through wonderfully poetic slogans such as 'the web of life', 'the wholeness of nature' or 'the cosmic oneness'. One of the major political aims of such views of the world is to promote environmental friendliness. Often pronouncements about the unity of nature are heralded as being so utterly important to humanity's ideas about itself and about nature that they have been described as being fundamental precepts within human knowledge.

The very idea of nature in unity is thought to stir the blood of humans towards a kindness to the environment. It is a concept that is supposed to inspire sensitivity within us; promoting a feeling of togetherness and belonging with the Earth. Unity is therefore something to be revered, cherished and respected.

This book, however, takes a different tack. Rather than simply revering the unity of nature, this work aims to investigate the intellectual milieu that has produced the concept. While doing so, I am driven to argue that the body of literature extolling this worldview of unity may well possess an internal coherence which superficially promotes environmentalism yet it is actually replete with a myriad of metaphysical, ethical and social connotations at odds with environmental preservation. Much of the discussion about the unity of nature has been made in an attempt to add value to, and protect, the environment but it is my contention that the unity of nature idea may actually do the reverse; it may devalue the environment and endanger it. This is most especially the case when the unity of nature idea makes knowledge claims with regards to the nature of the terrestrial animal and plant communities of the world. These ecological communities are, of course, prime sites of ecological value. However, I intend to suggest that such ecological value is undermined and overwhelmed when these ecological communities are made to submit to the 'unity of nature' idea.

A subtext that might be seen to emerge as a corollary to this critique is the laying down of an attempt to disenfranchise the ecological competence of many of the most well-known users of the unity of nature idea. Ecology is the science that 'unity of nature' supporters often appeal to and this is the science in which they believe they can have a lot of influence. It has become my view, however, that unitarian claims to know how to operate with, and extend, ecological concepts are riddled with mistaken assumptions and over-generalization.

The perilous connotations that emerge from the 'unity of nature' idea are not confined to the natural world; they are just as worrying for the social world too. Natural unity, we shall see, does not just exist on its own as an independent idea. It has a whole attendant army of supporting concepts, narratives and metaphors from which it gains its strength. For example, unity has attachments to the ideas of 'balance', 'order', 'hierarchy', 'stability', and the concept of the 'system'. In this book, these companion ideas are filtered out from the unity of nature idea and then, one by one, distilled so as to expose their unfortunate philosophical side-effects.

If I hold that the unity of nature might be an inappropriate worldview given the character of its contemporary expression, then I might reasonably be asked: 'What would a more appropriate ecological narrative look like?' My answer to this question, at least as it might apply to terrestrial ecological communities (most notably, forest communities), is foreshadowed throughout this work but reaches its climax in the final chapter where there is a discussion about a new way of looking at nature. This alternative ecological perspective is, I believe, just as well equipped to handle the ecopolitical needs of environmentalism as the unity of nature idea but it exists without needlessly inflicting the negative consequences of unity upon the social and ecological worlds.

When examining the intellectual milieu surrounding the unity of nature idea, an intense investigation of humanity's relationship with nature must ensue. Those who have explored this relationship are often deemed to be doing something called 'the philosophy of Nature'. If this is an apt term for a book such as this, it should be made absolutely clear that such a label should be interpreted in the widest sense. One does not have to be a philosopher to undertake studies in the philosophy of Nature, and--to be sure--sociologists, historians, scientists, theologians, environmental theorists, cultural theorists, art theorists, along with professional philosophers, have all studied various aspects of humanity's relationship with nature for many years. It is from all of these traditions that I harness various tools with which to examine the background and implications of the unity of nature.

Within these disciplines (and many more besides) there is an ongoing debate between naturalism and realism on the one hand and social constructionism on the other; and I should state at the outset that I am, myself, more allied to the latter than to the former. Having declared my constructionist sympathies, I can now admit that such a route aims to forego the production of the ultimate word on: 1) the nature of nature, 2) the nature of the social, 3) the objectivity of any explanation describing the interaction between the two. A constructionist approach nevertheless presents an opportunity to address some issues within environmental and ecological scholarship in a more creative, challenging and tangential way than so far accomplished. As we shall see in the final chapter, however, it is likely that environmentalism presents social constructionism with a challenging and tangential perspective, too. A theme explored in the last chapter, for example, suggests that social constructionism might need to anticipate its own widening to take account of a critique of humanity's hubristic claim to be the only social creatures upon the planet that are capable of social construction.

As an interdisciplinary investigation--one which traverses over a variety scholarly fields--it is likely that this work will, at times, wander through many contestable interpretations of standard theories that seem incredibly simplified to some readers and inordinately complex to others. Because my aim is to communicate to readers who hail from a wide variety of professions and with diverse interests, I am compelled to pitch the arguments at a level which confuses the least possible number of readers without falling into repetitively stating what most educated people might commonly believe to be obvious. To help me succeed in this endeavour, a glossary is included at the book's end so that specialised terms which may not be familiar to all can be referred to without continuously interrupting the reading with a lot of in depth explanations.

The unity of nature, because it is such a grand concept in contemporary thought, has a huge story behind it with many characters and many plots. So many, in fact, that the reader may feel giddy at one time or another as they are introduced to unfamiliar ideas and unfamiliar people. In light of this, it might be nice of me to introduce the main characters that I have used in my particular telling of the unity story. The main human characters are people like Edward Goldsmith, Fritjof Capra, Paul Davies and James Lovelock; philosophers of nature hailing from both science and environmentalism. Edward Goldsmith, for instance, is a well known British environmentalist and publisher who has philosophized aplenty about the unity of nature. Fritjof Capra is a best-selling physicist-author who has also investigated natural unity from a philosophical and scientific perspective. James Lovelock is the well-known inventor of both scientific gadgets and the Gaia theory; the latter of which has great repercussions for the unity of nature concept. Paul Davies is a world famous Australian-based scientist who has written numerous popular books that attempt to answer the big philosophical issues associated with the nature of nature.

Along with the works of these scholars, the works of other famous and not so famous characters of contemporary science and social science are investigated; including people like Lynn Margulis, the biologist who co-invented the Gaia theory; John von Neumann and Norbert Wiener, the inventors of a field of science called cybernetics, Eugene Odum, a 'founding-father' of ecology who popularized the ecosystem concept; Frederic Clements, and Henry Gleason, two early Twentieth century pioneers of ecological science; Talcott Parsons, the well-known mid-Twentieth century sociologist; Frederick Hayek, the Nobel prize winning economist, and Charles Birch, a theologian-philosopher whose ecotheological prescriptions have influenced numerous environmentalists for many years. By the way, if none, or only a very few, of these names are known to you, do not fear, their relevance to the themes explored in this book will be outlined as you read the story. The real star, though, is nature. And nature receives its most glittering applause and adoration when it is playing its unity role. If this is so then it is quite possible that everybody in Western society contributes to the unity of nature character since we have all played within its story at one time or another.

It should be noted here, in the introduction, that this work is organised into three parts which are divided into eight chapters. As an aid to the reader the chapters might be prefaced with the following comments:

Chapter One, 'Unity as an Environmental Idea: An Introduction to the Unity of Nature in Contemporary Thought', introduces the unity of nature as an identifiably environmental or ecopolitical worldview. It puts forward the uncontroversial notion that the current scientific conception of the unity of nature is, at least in part, a construction of the environmental movement. This particular social movement has much ecopolitical investment in the idea. To environmentalists, the unity of nature suggests the need to tread lightly on the Earth since what we may do to one part of it will always have ramifications for many other parts.

There is a common historical narrative told and retold by environmentalists when they discuss the broader cultural uptake of these views. This common historical narrative is recounted in Chapter One and it is a narrative that generally suggests that bad ideas, especially those which posit a non-united nature, lead to bad environmental practices. From this perspective the environmental crisis is, in some way, a result of an ecopolitically inappropriate worldview.

Chapter Two, 'Falling Into Wholes: Ecological Fascism', makes the point that, despite their best intentions, unitarians have often entered into an ecological and environmental discourse which exhibits an over-zealous use of holism. This over-zealous use of holism is described in this book as 'ecological fascism'; where the value of individual living beings are discarded for the good of the whole.

Chapter Three, 'Gaia: The Technocentric Embodiment of the Unity of Nature', addresses one of the best organised attempts at establishing a distinct theory of natural unity: the Gaia theory. As a brand of unitarianism that is intellectually tied to those analysed in Chapter Two, it is suggested that the Gaia theory might not be as ecopolitically relevant as some people have thought. Certainly, it is premature to proffer Gaia theory as the new environmental worldview for the Twenty First century as some environmental thinkers have done. Gaia, in fact, may just be the product of an overly technological approach to ecological and environmental problems.

Chapter Four, 'Natural Conservatism: The Unity of Nature and Social Systems', examines whether or not the unity of nature concept might possess conservative political tendencies when it is transferred from the study of nature and utilised in the study of society. The conclusion reached in this book is that it does. Whether or not these tendencies are inherent and necessary is still left open but the specific intellectual milieu surrounding and feeding into the particular unity ideas that environmentalists and ecologists promote is profoundly continuous with the unity ideas promoted by conservative political thinkers.

Chapter Five, 'Uniting the Ecosystem with the Economy', examines the links between the unity of nature concept and its recent scientific representation within new scientific theories such as Chaos theory, complexity theory, and self-organisation theory. It then investigates the links between these theories and the philosophical ideas of liberal capitalism. Chapter Five suggests that these newest configurations of unity are intellectually and historically associated with some of the oldest self-legitimising ideas within liberal capitalism. If environmental thinkers are to continue to expound the idea of unity in their philosophical treatments of nature, they may find that they are expounding supportive arguments for liberal capitalism also.

Chapter Six, 'What Is This Thing Called Postmodern Science?', investigates how the unity of nature concept has become a central tenet in Postmodern Science and then goes on to identify the nature of Postmodern Science and its relationship to the general postmodern intellectual movement. In this chapter there is a discussion about how Postmodern Science is unable to rid itself of many Modernist tendencies. Because of this, it is concluded that Postmodern Science, despite its name, is not really postmodern.

Chapter Seven, 'Mechanicism Vs Organicism: A False Dichotomy?', looks at the links between the unity of nature, mechanicism, and organicism. It then goes on to indicate how these three things are intimately connected. Their connections are such that it no longer seems tenable to regard mechanicism and organicism as generic polar opposites in environmental and ecological thinking.

Chapter Eight, 'An (Other) Postmodern Ecology', assumes that Chapters Two through Seven have shown the inadequacy of the unity of nature idea as an environmental and ecological view of the world. Chapter Eight seeks to investigate the possibility of suggesting an alternative postmodern conceptualisation of ecology whereby nature itself may be deconstructed in an ecopolitically benign way through an awareness of 'Otherness'. Within such a conceptualisation it is held that we should not be scared of identifying and locating the differences between the various members of the environment; splitting them up and atomising them, since if we do it with a view to celebrating Otherness then we may well contribute to an environmental narrative that is better equipped to value the individual lives of each living member in the world.

These chapters are arranged into three parts which delineate the peculiar intellectual themes to which they belong. The first part *(It's All an Environmentalist Plot!)* suggests how environmental thinkers have constructed the unity of nature idea as part of a contemporary environmental view of nature. Part Two (*It's All a Bourgeois Plot!*) revolves around the consideration that the unity of nature is just as much a social construction of conservatism, fascism and liberal capitalism as it is a construction of environmentalism. Part Three *(It's All a Postmodern Plot[lessness]!)* develops the theme that the unity of nature idea may be a decidedly postmodern concept and it investigates how 'constructive' Postmodern Science constructs the unity of nature and how 'deconstructive' postmodernists might deconstruct it.

PART A

It's all an Environmentalist Plot!

Part A recounts the commonly accepted idea that the unity of nature concept is, essentially, an environmental idea. Environmentalists have colonised the unity concept, reinvented it, and reinvigorated it to serve as a legitimising tool for their ecopolitical goals. This colonisation, reinvention and reinvigoration has been so successful that, nowadays, the unity of nature is commonly accepted within many facets of Western culture as the intellectual preserve of environmentalists, ecologists and their sympathisers. Part A contains only one chapter: 'Unity as an Environmental Idea: An Introduction to the Unity of Nature in Contemporary Thought'. This chapter serves as an introduction to the rest of the chapters in this work. It is important since it lays down, as a baseline, the fundamental partnership between environmentalism and the 'unity of nature' idea as conceived by its environmental supporters. It is important in another respect since it also lays down the fundamental relationship between scientific thinking and social practice as conceived by both environmental thinkers and many philosophers of Nature. It also recounts a peculiarly environmental narrative with regards to the history of scientific thought; a narrative which divides the history of Western philosophy of nature into three distinct stages whilst proposing that early Twenty First century society sits at the cusp of the second and third stages.

CHAPTER 1

Unity as an Environmental Idea:
An Introduction to the Unity of Nature in Contemporary Thought

Thinking About Nature: Why Is It Important?

Social paradigms are linked to scientific paradigms. This is the thesis of an enormous number of well-known workers in environmental thought and the philosophy of Nature. For example a short list of environmental thinkers who would adhere to this thesis might include Fritjof Capra, Edward Goldsmith, Bill Devall, George Sessions, David Suzuki, Jonathon Porritt, Theodore Roszak, Dave Foreman, Frederick Ferre, Carolyn Merchant, David Abram, David Bohm, John Cobb, Charles Birch, David Pepper and David Griffin. A longer list of authors might include countless others in both academic and non-academic circles who write about science, nature and the environment.

Within such a framework it is held that scientific outlooks, metaphysical overviews, or cosmological worldviews--call them what you will--profoundly affect social practices and political activities. By extending historian Thomas Kuhn's notion of scientific paradigms[1] into the wider realm of the social, environmental thinkers and philosophers of nature have made links between how we think about nature and the effect of this thinking on both nature and society. By doing this, environmental thinkers are able to observe which philosophies and worldviews give rise to which social and ecological situations. In environmental thought it is apparent that very specific ideas seem to have given rise to very specific social practices.

With regard to this last point it is notable that environmental scholars often lay much of the blame for the environmental woes of the past century on the fragmented, mechanistic and atomistic way that the natural environment is viewed since this way of viewing the world produces a fragmented, partial and incomplete appraisal of environmental problems and fosters a fragmented, partial and incomplete set of practices that might steer us away from the environmental crisis.

What is needed, environmentalists generally say, is a recognition of, and a commitment to, an age-old truism: the natural world exists as an interconnected unity. Or in the words of the true believers:

> We must rediscover the unity of humanity and all creation.[2]
>
> The grandeur and majesty of oneness I have only found in nature.[3]
>
> Life is fundamentally one.[4]
>
> No one can deny that we are all interdependent with each other and with all of nature.[5]
>
> The biosphere is one.[6]
>
> To accept the biocentric outlook and regard ourselves and our place in the world from its perspective is to see the whole natural order of the Earth's biosphere as a complex but unified web of interconnected organisms, objects and events.[7]

When suggesting and reiterating the unity of nature, the writers above are making political as well as conceptual statements. Environmentalists are using unity to claim that nature can only be looked after properly if it is held to be united. Within this structure of thinking, humans are held to be morally obliged to look after nature because they are a part of it. If humans are held to be separate from nature, or if nature is held to be separated into various bits and pieces, then environmental problems will continue indefinitely since humanity will not be obliged to care for something that is not connected to itself (i.e. that does not affect humans) and its members will always presume that what they do to one part of nature will only be localised to that part (and therefore they will hardly worry about the affect it has on the rest of nature). These ideas are not controversial: they are merely reflecting the mainstream environmental line when dealing with philosophical matters about the nature of nature.

If we are convinced of the idea that how we think about nature somehow determines how humanity operates in practice then it is obvious that how we think about the world is profoundly important. Within environmental writing there is a lot of discussion which tries to detail the precise relationship between how we think about nature and how this thinking affects both ourselves and nature. Here are a few examples:

In this phase of human history there is widespread conflict between our conception of ourselves and our conception of the world. We see ourselves as beings that are conscious, that are rational, we have free will and are purposive. But we see the world as consisting of mindless, meaningless, totally determined physical bits and pieces that are non-purposive. A society that lives with this dichotomy is operating on a profound error that is destroying much that is worthwhile both in ourselves and in the world.[8]

The major problems of our time--the growing threat of nuclear war, the devastation of our natural environment, our inability to deal with poverty and starvation around the world, to name just the most urgent ones--are all different facets of one single crisis, which is essentially a crisis of perception...It derives from the fact that most of us, and especially our large social institutions, subscribe to the concepts of an outdated worldview, inadequate for dealing with the problems of an overpopulated, globally interconnected world.[9]

We have seen that fragmentary thinking is giving rise to a reality that is constantly breaking up into disorderly, disharmonious and partial, destructive activities.[10]

In these quotes a specific link is commonly alluded to: bad thinking, especially that which posits a fragmentary nature, leads to bad environmental practices. The environmental crisis is thus, at least in part, a result of 'outdated', 'erroneous', 'fragmentary' philosophies of nature.

The Story So Far: The History of Scientific Worldviews According to 'New Paradigmers'

During their elaboration of the 'bad thinking--bad practices' link, environmentally-inspired scholars have often embarked upon a three stage history of the origins and development of this situation. It goes something like this:

Stage 1: Ancient Philosophies of Nature:

Before modern times the world was not viewed as 'fragmented' and 'dead' but 'organic' and 'alive'. Humans did not regard themselves as over and above nature but as an intimate part of it. This worldview has sometimes been labeled the 'organic' worldview.[11] According to the modern day physicist and philosopher of nature; Paul Davies, it was epitomised in ancient times by:

> The Greek philosopher Aristotle [who] constructed a picture of the universe closely in accord with this intuitive feeling of holistic harmony.[12]

This organic view of the world characterised by Aristotle (and more importantly, according to George Sessions, by the pre-Socratics[13]) vitalised the Earth, describing all of its members as alive rather than dead and interconnected rather than isolated. The Earth, itself, was also seen as alive. It possessed an overall harmony and integrity much the same as the harmony and integrity of a single living creature. Such respect for, and emphasis on, living things, and upon the unity of all these living things, suggests, according to modern day environmentalists, a non-anthropocentric value system since humans were living in a world with other beings whom they were deeply interconnected with, and they were also living in a world where the Earth itself was regarded as an organic being.

Ancient Western philosophy was not the only intellectual site for such organic worldviews and much is said within environmentalism about how traditional and non-Western cultures had, almost invariably, believed in the sacred unity of the world and the sacredness of non-human beings within it. For such traditional cultures, the sacred unity of nature bestowed a certain moral and cosmic outlook:

> These cosmologies, involving a sacred sense of the Earth and all its inhabitants, helped order their lives and determine their values.[14]

The people within traditional societies lived their lives and cultured their values in order to revere both the world and the beings it.

Stage 2: Cartesian philosophy and the advent of the Modern Worldview.

Enter Rene Descartes, the environmental evil one; mid-wife of mechanicism and hinderer of holism. Within environmental thought, Cartesian philosophy is now usually held to have "separated mind from body"[15], isolating the world from the observer. Cartesian thought emphasised the mechanical nature of things; thus breaking up the unity and interconnectedness of the universe into machine-like parts. According to Danah Zohar this:

> mechanicism stresses an unbridgeable gulf between human beings and the physical world...We can see in such a distorted perception the origins of our current ecological crisis.[16]

A break was thus made between ancient/traditional worldviews and the modern scientific worldview, and also between traditional and modern ways of looking at life and living things:

> Whereas traditional biology had explained the activity, organisation, and direction of life by a hierarchy of vegetative, sensitive and intelligent faculties or souls, Descartes reduced the involuntary physiological process of the body to mechanical processes and eliminated the vegetative and animal souls from physiology. Sensitivity and thought were confined to humans and attributed to an immaterial rational soul.[17]

As the modern mechanical worldview took hold there was, according to Paul Davies, an:

> emphatic rejection of all forms of animism or finalism, and most modern biologists are strongly mechanistic and reductionist in their approach.[18]

Or as David Pepper explains: "Descartes extended the concept of nature as a machine initiated by Kepler and Galileo"[19] so that Cartesian philosophy:

> viewed animals, and the human body, too, as machines. They were automata, and their workings could be fully known by reducing them to matters of physics and chemistry, which in turn could be understood in terms of mathematics.[20]

All in all, then, Cartesianism was the onset of the 'Modern Era', and:

> with the coming of the Modern era...nature could be thoroughly understood and eventually brought under control by means of the systematic development of scientific knowledge.[21]

In the Modern era "Aristotelian ideas were banished from the physical sciences"[22] while purpose and teleology as explanatory devices were confined to the intellectual dustbins of superstition and religion. The modern era "pictured the universe as a gigantic contrivance and ourselves as small contrivances or machines".[23]

Environmentalist versions of the history of scientific thought generally continue in this vein with regards to the rise of mechanicism in the Seventeenth century where they castigate Isaac Newton as a co-conspirator against nature: "Newtonian physics was in the same spirit as Cartesian philosophy"[24] it is said, and "the general picture most of us have about the world is derived from Newton's mechanics of the Seventeenth century".[25]

To those who tell the story, Newtonian physics rejected natural freedoms in favor of physical determinism, it converted all things to machines, devaluing them in the process, before going on to simplify the varied processes of nature into simple mechanical systems of interaction like the physics of a billiard-ball table:

> In classical Newtonian mechanics, once the initial conditions and the force laws are given, everything is calculable forever before and after. The system is governed completely by the laws of mechanics and of conservation of energy. It is totally determined. It has no freedom.[26]

As time went by, and as science developed, things only got worse. According to David Bohm:

> the mechanistic view in physics...was a characteristic of the Modern view and... reached its highest point towards the end of the Nineteenth century...this view remains the basis of the approach of most physicists and other scientists today.[27]

When scientists, and science popularisers, in the mid-Nineteenth century became aware of their own power to prescribe worldviews for the rest of society to take notice of, and when these people rallied with the Western world's leaders and managers to promote an optimistic view of relentless scientific and technological progress, then a new belief of inevitable onward human betterment took hold.

The Modern paradigm forged by Descartes and Newton, and self-consciously disseminated by Nineteenth century scientists and philosophers, promoted a picture of a dead, fragmentary, machine-like world. This picture, environmentalists think, was entirely inimical to nature and her members; ushering in a world devoid of value and spirit, a world which could be manipulated and changed with gay abandon, a world where materialism and industrialism could become rampant, a world where nature could, ultimately, be transformed into a mere resource for humanity's various whims and desires.

When trying to unearth the negative effects of the mechanical worldview it is often the case that three philosophical relatives of mechanicism receive blame as well. These relatives are 'reductionism', 'atomism' and 'dualism'. As Davies points out reductionism is:

> The procedure of breaking down physical systems into their elementary components and looking for an explanation of their behaviour.[28]

Reductionism is usually thought to be strongly related to atomism, a view that all the world is made up of distinct, independent bits and pieces:

> The Newtonian paradigm fits in well with the philosophy of atomism...The behaviour of a macroscopic body can be reduced to the motion of its constituent atoms moving according to Newton's physical laws.[29]

As Paul Davies and Charles Birch would explain it, these law-obeying bits and pieces are generally what reductionism reduces natural phenomena to. In Modern science such bits and pieces might be atomic or sub-atomic particles or they might be the molecules that make up the genetic codes of living things.

Dualism, the third relative of mechanicism, is variously held to be that brand of Cartesian philosophy that either separates 'mind from matter' or 'humans from non-humans'. Since, within Cartesian thought, humans are the only things that possess 'mind' then both of these variants amount to the ejection of non-humans from the realm of the truly conscious and animated. Non-human entities, including our own human bodies, are mere machines with no animate soul. The human mind, on the other hand, exists above these things as the divine or evolutionary epitome of all existence since it is the realm to which all thinking and feeling belongs.

According to many modern day environmental scholars, this quartet of Modernist foes; mechanicism, dualism, reductionism and atomism, are the intellectual harbingers of death. They kill the environment by reducing the natural world to broken up bits and pieces which have no soul, no animation, no life. The only thing within this world that does possesses life in its full and glorious extent, and therefore the only thing worthy of complete respect, is human consciousness--and as such everything else is subordinate in value to it.

You do not have to be an environmentalist to see mechanicism, dualism, reductionism and atomism as the quartet of death. For contemporary philosophers of nature who have no special interest in the environmental crisis, like Paul Davies for example, this quartet of death has manipulated science and produced a worldview of morbid decay and pessimism. Davies admits that mechanicism has succeeded in providing modern society with very many of its scientific and technological marvels but he feels that these have come at the cost of giving rise to a world seen as meaningless and purposeless.[30]

This modern tendency towards meaninglessness and purposelessness is most stark, according to Davies, if we consider the grand cosmological implications of the mechanical worldview. Once the mechanistic worldview was interpreted through the Nineteenth century science of thermodynamics and then reapplied to the universe as a whole, it was soon realised that the universe was decaying. Just as a Nineteenth century machine could not hope to operate without converting some of its fuel into waste heat as it struggled to do productive work, so the universe as a whole must struggle with its daily loses of energy through heat waste until one day it has no energy left to exist. Or as Davies puts it:

> Everyday the universe depletes its stock of available, potent energy, dissipating it into waste heat. The inexorable squandering of this finite and irretrievable resource implies that the universe is slowly but surely dying.[31]

Due to the cosmological picture that Newtonianism presents to modern science and modern society, humanity has come to believe "that the total useful, or working, energy of the universe, according to the laws of physics,...is gradually running down."[32]

If the universe is dying--slowly decaying towards non-existence--then, alludes Davies and many others, a vast shadow of pessimism seems to descend upon all human endeavours. Any hope for an eternal life for humanity, or just a better one for our descendants, is pointless since the universe itself will one day die, taking with it any degree of purpose that may have existed with it.

Stage 3: The New Paradigm:

Because "there is growing dissatisfaction with sweeping reductionism, a feeling that the whole really is the sum of its parts", and because of "the sweeping nature of scientific discoveries in cosmology, fundamental physics and biology,"[33] Paul Davies, amongst many others, believes that a new set of scientific disciplines are arising which present a worldview of the cosmos somewhat reminiscent of the pre-modern cosmology of Stage 1. Of this reassertion of the traditional way of thinking about nature, Charles Birch says:

> The mechanical images no longer fit. They are giving way to quite a different image of the universe and ourselves. This discovery is being made simultaneously by a science, a philosophy and a theology as yet little known. Its new images are no longer mechanical: they are organic and ecological. The universe turns out to be less like a machine and more like a life. This constitutes a new revolution in the science, philosophy and theology of our time.[34]

Fritjof Capra, too, expresses the eclipse of the modern way of thinking in similar terms:

> The social paradigm now receding has dominated our culture for several hundred years, during which it has shaped our modern Western society and has significantly influenced the rest of the world. The paradigm consists of a number of ideas and values, among them the view of the universe as a mechanical system composed of elementary building blocks, the view of the human body as a machine, the view of life in society as a competitive struggle for existence, the belief in unlimited material progress to be achieved through economic and technical growth.[35]

Of the new way of thinking, Birch and Capra declare that a more complex, life-affirming, interactive and interdependent sensibility is emerging; one which affirms the principles of ecology and reaffirms the ancient ideas of unity and holism. The new paradigm associates the world with living things rather than with machinery, with complex integrated organisms rather than with simple mechanical entities that can be broken up.

The new emerging paradigm, in contrast to its predecessor, reasserts the organismic, biological basis of reality in an Aristotelian vein. The living organism becomes the metaphor of choice and nature is no longer regarded as a disparate collection of clashing parts but as an interconnected and interdependent unity. Instead of a worldview of mechanicism, reductionism and atomism, the new paradigm advocates the world to be organic and holistic.

Paul Davies and Arthur Fabel find that intellectual and cultural palatability of the new emerging worldview revolves around its inherent optimism since while Cartesianism and Newtonianism promoted a worldview of death and decay for the universe, the new paradigm "just now arising, may set aside this pessimistic view, replacing it with a conception of the cosmos as a self-organising genesis."[36] Davies, in this regard believes that the social acceptability of holism is a reaction against the pessimism and "despair of a reductionist universe."[37]

Fritjof Capra, Edward Goldsmith, Charles Birch, David Bohm, Donald Griffin, Charlene Spretnak and many others, however, feel that the social repercussions go much further than this; that the new paradigm is producing new environmentally friendly narratives of the living world: "the emerging new paradigm may be called a holistic or an ecological, worldview"[38], and by:

> calling the emerging new vision of reality 'ecological' ...we emphasise that life is at its very centre. This is an important point for science, because in the old paradigm, physics has been the model and source of all metaphors for all other sciences.[39]

Armed with the new worldview, or with a re-emergence of the ancient worldview (whichever way you want to look at it) many of the New Paradigmers have conjured up grand ideas about their ability to now go ahead and change the world. After all, if their basic assumption that scientific paradigms can influence social paradigms is correct, then there may be no stopping them from influencing the rise of new regimes of social organisation based on their new worldview:

> Researchers in several scientific disciplines, various social movements, and numerous alternative organisations and networks are developing a new vision of reality that will form the basis of our future technologies, economic systems and social institutions.[40]

Technical Prelude to Parts B and C

This three-staged historical narrative is common throughout the body of literature that has amalgamated itself under several titles such as 'the New Paradigm' or the 'the organic worldview'. It is a story which is told repeatedly within this body of literature.[41]

There are (for scholars from various disciplines) a number of obvious problems with this three-stage environmental version of the history and philosophy of changing 'paradigms'. Most notably:

1) are the links between social and scientific paradigms really as the new paradigmers describe? (i.e. does bad thinking really give rise to bad practices)?

2) if we admit that bad thinking can lead to bad practices, what relevance has this got to environmentalism? (i.e.: does this necessarily mean that bad environmental thinking has caused the environmental crisis)?

3) is the representation of Cartesian/Newtonian/Modernist philosophy as contributory historical factors in the rise of anti-environmental thought entirely valid? (i.e.: might the environmentalist characterisation and periodisation of Cartesianism and Newtonianism be less than solid bases upon which to make comparisons of pro-environmental and anti-environmental concepts)?

These problems are important but they are not the ones considered in the chapters that follow. Instead this work forthrightly strides into the arguments that the New Paradigmers themselves make; entering into the terms of the debate as actually set out by the environmental thinkers and philosophers of nature that adhere to the New Paradigm of unity, organicism and holism, rather than stepping back and attacking them from outside their own intellectual domain. In doing this, however, we see that the terms of debate internal to the philosophies of the New Paradigm are often flawed (in the way they philosophise and politicise natural phenomenon) and also self-contradictory (in that the New Paradigm is riddled with the very social and environmental dangers that New Paradigmers accuse Modern Paradigmers of). These flaws and contradictions will be explored throughout the rest of this book.

Before I go on beyond this first chapter, however, a warning must be issued with regard to some labels. Where I have talked of 'New Paradigmers' here as those that hold to the philosophical beliefs of organicism, holism and unity, in other chapters (and for specific reasons) they come to be known under various other labels. For instance, in Chapters Two to Four I refer to them under the general name of 'unitarians' and 'holists', in Chapter Five they are referred to as 'New Scientists', and in Chapter Six they become known as 'Postmodern Scientists'. There are good reasons for me to use these various labels at varying points in this book since I am shadowing the self-conscious invention of these names by their users who, despite slight variations, can all be described as followers of the unity idea. I shall alert the reader of the reasons for these various name changes as the book proceeds. Note that I shall only endeavour to jump from one name to another after adequate warnings as to the exact relationship between those who operate under the respective names.

As far as labels like 'idea', 'concept', 'paradigm', 'worldview' and 'metaphysics' go, I generally use these words in an overlapping manner because, again, I shadow the terminology as played out in the writings of those philosophers of nature whom I examine. For instance, Paul Davies usually sticks to the label 'worldview' to describe his grand schemes about the nature of the universe, whilst someone like Donald Griffin utilises the label 'paradigm' for the same sort of job.

Despite the above point, some readers might like more precise boundaries with regards to the way I use such labels. If this is so, then these readers might take the label 'concept' to describe an abstract idea whose abstractness may not be generally acknowledged within the literature, while I utilise the word 'metaphysics' to describe an abstract idea whose abstractness is well-known at least in the scholarly literature. The labels 'paradigm' and 'worldview' I take to refer to a concept or a metaphysical point of view that has become so general within a certain scientific or intellectual framework that it is used to categorise and explain all (or many) other concepts. Another name for such all-embracing stories is 'metanarrative', and this is a word (derived from Jean-Francois Lyotard's work[42]) that is sometimes favoured when discussing the story of unity since it is the more favoured contemporary scholarly term when describing grandiose intellectual theories.

What I intend all this to mean is that when Griffin talks of a 'paradigm'; when Davies talks of a 'worldview'; and when Lyotard talks of a 'metanarrative', they are all referring to a grand story or idea that is held to possess such undeniable explanatory ability that it forms the basis of a whole field of knowledge.

Whilst explaining labels I might also introduce some terms that are sometimes used synonymously in the literature that I am exploring but which I, myself, tend to differentiate within this book. In this particular work 'ecology' refers to the science of ecology and not to the social movement involved with protecting specific ecological settings. The latter is termed 'environmentalism'. A contributing factor to the prevalence of the notion that ecology deals with and operates within the paradigm of natural unity is this conflation between ecology and environmentalism under the term 'ecology'. Often people hear talk coming from 'ecologists' who are actually 'environmentalists' and then go on to suppose that their messages (including those of metaphysical unity) are part of an acknowledged scientific tradition and/or position. Most, but not all, writers examined in this work, do not succumb to this conflation but I suggest that their readers often do.

There is one more regularly used label that might need clarification in this first chapter; the label 'philosophy'. Some readers, notably philosophers, might have a problem with the way I bandy about the name of a scholarly tradition with grave carelessness. They might note, for instance, that what the many 'philosophers of nature' named in this work are doing is not actually 'philosophy' but some inferior brand of cogitation which deserves to go by another name. It should be noted, however, that I am using 'philosophy' in its most liberal and popular sense; to describe all the big ideas, great concepts, profound thoughts, worldviews and outlooks on life that all people (philosophers, scientists, environmentalists, theologians, both amateur and professional) are capable of exploring.

Notes to Chapter 1 (Unity as an Environmental Idea)

1. See T.S. Kuhn, *The Structure of Scientific Revolutions* (Chicago: University of Chicago Press, 1996, originally printed 1962).
2. J. Porritt, *Seeing Green* (Oxford: Blackwell, 1984: 211)
3. C. Spretnak, *The Spiritual Dimension of Green Politics* (Sante Fe: Bear, 1986).
4. A. Naess, *Ecology, Community, Lifestyle* (Cambridge, UK: Cambridge University Press, 1989) p.192
5. S. Nicholson.(1992) "The Living Cosmos", in S. Nicholson and B. Rosen (eds*) Gaia's Hidden Life* (Wheaton, Ill. Quest Books, 1992) p.11.
6. E. Goldsmith, *The Way: An Ecological Worldview* (NY: Shambhala, 1993), p96.
7. P.W. Taylor, "The Ethics of Respect for Nature", *Environmental Ethics*, 3, 1981, p.211
8. C. Birch, *On Purpose* (Sydney: NSW University Press, 1990) p.ix
9. F. Capra, *The Web of Life* (HarperCollins, NY: 1996) p.4.
10. D. Bohm, (1994) "Postmodern Science and a Postmodern World", in C. Merchant, ed, *Key Concepts in Critical Theory: Ecology* (NJ: Humanities Press) p350.

11. For example by most of the above named authors, like Goldsmith, Spretnak, Nicholson and Merchant.
12. P. Davies, *The Cosmic Blueprint* (Harmondsworth: Penguin, 1987) p.6.
13. George Sessions points out that describing Aristotle as some sort of ancient environmental hero is somewhat dubious since "although Aristotle's philosophy was biologically inspired, nevertheless he arrived at a hierarchical concept of the 'great chain of being' in which nature made plants for animals, and animals were made for the sake of humans." See G. Sessions, "Ecocentrism and the Anthropocentric Detour", in C. Merchant, ed, *Key Concepts in Critical Theory: Ecology* (NJ: Humanities Press, 1994) p142.
14. *ibid*, p140.
15. D. Zohar and I. Marshall, *The Quantum Society: Mind, Physics and a New Social Vision*, (London: Flamingo, 1993) p.46.
16. *ibid*, p.5.
17. P. Christensen, "Fire, Motion, and Productivity: the Proto-Energetics of Nature and Economy in Francois Quesney", in P. Mirowski, ed, *Natural Images in Economic Thought* (Cambridge: Cambridge University Press, 1994) p.252.
18. Davies, *The Cosmic Blueprint*, p.8.
19. D. Pepper, *The Roots of Modern Environmentalism* (Beckenham, UK: Croon Helm, 1984) p50
20. *ibid*.
21. Bohm, "Postmodern Science and a Postmodern World," p.432.
22. Davies, *The Cosmic Blueprint*, p.8
23 Birch, *On Purpose*, p.x.
24. Zohar and Marshall, *The Quantum Society*, p.108.
25. Birch, *On Purpose*, p.ix.
26. *ibid*.
27. Bohm, Postmodern Science and a Postmodern World," p.343
28. Davies, *The Cosmic Blueprint*, p.13
29. *ibid*.
30. This can be observed in Davies' *The Cosmic Blueprint* and in P. Davies, *The Mind of God* (London: Penguin, 1993)
31. Davies, *The Cosmic Blueprint*, p19
32. E. Sahtouris, "The Dance of Life", in S. Nicholson & B. Rosen, eds, *Gaia's Hidden Life* (Wheaton, Ill.; Quest Books, 1992) p.21.
33. Davies, *The Cosmic Blueprint*, p.8.
34. Birch, *On Purpose*, p.xi.
35. Capra, *The Web of Life*, p.6.
36. A.J. Fabel, "Environmental Ethics and the Question of Cosmic Purpose", *Environmental Ethics*, 1994, 16, p.303.
37. Davies, *The Cosmic Blueprint*, p.197)
38. Capra, *The Web of Life*, p6.
39. *ibid*, p.12.
40. F. Capra, "Systems Theory and the New Paradigm", in C. Merchant, ed, *Key Concepts Critical Theory: Ecology* (NJ: Humanities Press, 1994) p.335).

41. Apart from being told in the above named publications the story recounted in this chapter is alluded to or retold in countless other publications to do with the history and philosophy of the environment. A short list of such publications might include: T. Roszak, *Person/Planet: the Creative Disintegration of Industrial Society*, (London: Victor Gollancz, 1979); C. Merchant, *The Death of Nature: Women, Ecology and the Scientific Revolution* (San Francisco: Harper & Row, 1980); F. Capra, *The Turning Point* (NY: Simon & Schuster, 1982); J.B. Cobb, (1988) "Ecology, Science and Religion: Toward a Postmodern Worldview," in D.R. Griffin, ed, *The Reenchantment of Science* (Albany: SUNY Press, 1988) pp.99-115; B. Devall and G. Sessions, *Deep Ecology: Living As if the Earth Really Mattered* (Layton: Gibbs M. Smith, 1985), D.R. Griffin, ed, *The Reenchantment of Science* (Albany: SUNY Press, 1988); T. Roszak, *The Voice of the Earth* (NY: Simon and Schuster, 1991); D. Abram (1992) "The Mechanical and the Organic: On the Impact of Metaphor in Science", *Wild Earth*, 2, 2, 70-75, F. Ferre, *Hellfire and Lightning Rods: Liberating Science, Technology and Religion* (NY: Orbis Books, 1993); A. Tilby, *Science and the Soul* (London: SPCK, 1993), A. Lemkow, *The Wholeness Principle: Dynamics of Unity Within Science, Religion, and Society* (Wheaton, IL: Quest Books, Revised ed. 1995); A. Gare, *Postmodernism and the Environmental Crisis* (London: Routledge, 1995), C.M Bache and S. Grof, *Dark Night, Early Dawn: Steps to a Deep Ecology of Mind* (Albany: SUNY Press, 2000), D. Landis Barnhill and R.S. Gottlieb, *Deep Ecology and World Religions* (Albany: SUNY Press, 2001), R. Sheldrake *et al*, *Chaos, Creativity and Cosmic Consciousness* (Park Street Press, 2001).
42. J-F. Lyotard, *The Postmodern Condition* (Manchester: Manchester University Press, 1984).

PART B:

It's all a Bourgeois Plot!

Part B rewrites the story of the unity of nature to differ from that recounted in Part A. Instead of suggesting that the unity of nature is a construction of environmentalism this part of the book suggests that the unity of nature idea is the construction of potent conservative, fascist, capitalist and technocentric forces in contemporary Western culture. The unity of nature could thus be described by those of a left-wing persuasion as a product of the bourgeoisie; the product of those with capital and power.

Part B is divided into four chapters. Chapter Two looks at some ecological problems within unitarian thought, examining the possibility that contemporary renderings of the unity of nature give rise to a type of holism which is ecologically fascist since it denies the value and stories of individual organisms in favour of collective wholes. Chapter Three investigates the philosophical and environmental premises of the Gaia theory and how it, too, is affected by the same type of holism described in Chapter Two. Chapter Three also examines some of the technocentric and anthropocentric impulses within the Gaia theory. Chapter Four works with the familiar sociological notion that biological intrusions into the social sciences are often of a conservative bent. This notion is examined in relation to the unity of nature. Chapter Five identifies yet another distinct intellectual impulse within the unity of nature; its close heritage to liberal capitalist thought. This heritage flows into contemporary versions of the unity of nature idea via various political, philosophical and metaphysical conduits. These conduits are studied so as to betray their association with the liberalist (neo-classical) agenda in contemporary Western economics.

What tends to unite all of these chapters together is a critical focus upon the concept of the 'system'. The system concept has become an enormously popular concept within the minds of those with environmental sympathies, due primarily to the way it has been seen to tie nature and humanity together. Thus, any critical evaluation of the unity of nature idea must look at the use that is made of the system concept, especially when so many philosophers of nature readily assume that the processes going on within Earth's natural ecological systems are held to give enormous insight into the operation of human social systems.

The invasion of biological models into the study of social phenomena has often been a cause of concern (for example, see the work of Marshall Sahlins, Robert Young and Levins and Lewontin), and as noted above, many social scientists see biological explanations of social behaviour as being intrinsically conservative. At this introduction to the following chapters, I should indicate that I do not necessarily adhere to this criticism, though (as reflected in Chapter Four) I do have great sympathy for it. Rather, I adhere to a view that says that we cannot get away from narrating about the social world without the use of narratives which at one time or another were used to narrate about nature. A social narrative must not be castigated as inappropriate just because it seems to come from nature to society. Each narrative has to be judged on its own merits, within the historical and political context of its use. The important thing to note here, as I introduce the chapters of Part B, is that unity of nature is a narrative incapable of providing a value framework that works consistently well enough within ecopolitical discourse.

CHAPTER 2

Falling Into Wholes:
Ecological Fascism

Introduction

As noted in Chapter One, exclamations about the natural reality of unity seem to be a theoretical and rhetorical necessity for many environmental thinkers. All is not well, or resolved, in this regard however. By placing an over-arching emphasis upon unity, environmental thinkers might open themselves up to the criticism of being branded promotional agents for some politically offensive ideas that they would normally, themselves, protest against. Ecological fascism is one particular charge that surfaces in this regard and this chapter will explore this charge.

The sense in which I use the word 'fascist' revolves around the tendency to devalue individuality in favour of totality while being intolerant of plurality, difference and dissension. It is not a term that I utilise to act as an inflammatory metaphor but an indication that unitarians tend to glorify the ecological totality over the interest of individual beings. The reader may think this has only a very vague reference to fascist politics, and this might be so, but there is a certain resonance when we remember that "the term fascism is derived from the 'fasces' of ancient Rome, a bundle of rods with a projecting axe symbolising unity and authority."[1]

Ecological Fascism

If the emphasis is upon unity and holism in the natural world then does not the individual (whether human or not) get fascistically swallowed up in the whole unity? According to many unitarian thinkers, it does not. For instance Joanna Macy states:

> Do not think that to broaden the construct of the self...involves an eclipse of one's distinctiveness. Do not think that you lose your identity like a drop in the ocean merging into the oneness of Brahman. From the systems perspective, this interaction, creating larger wholes and patterns, fosters and requires diversity. You become more yourself. Integration and differentiation go hand in hand.[2]

Alas these retorts in the face of the fascism charge do not allay the fears of those making the charges. D.E. Marietta signifies this when she states:

> The dire picture of the effects of holism is not simply a response to what the holists have said. It is also based upon possibilities that holists did not mention and that they may not have foreseen.[3]

Any claim that a worldview of natural unity might immediately give rise to dangerous fascist ideas within social thought can only come about if there is slippage between ecological unity and social unity. This slippage is made possible in environmental thinking, as it is in a lot of scientific thinking, by the (albeit sometimes vague) acknowledgement that there is an ontological unity between humanity and nature. As humans are biological, ecological and physical organisms, environmentalists generally claim that there is no distinction between humans and the rest of the ecological world. We humans are derived from that world (both historically and physically) and we are an intimate interacting part of it. Many people have thus drawn the conclusion that the natural world in a strong sense shapes human nature and human society; or to say this differently: society reflects at least some natural processes.

However, others that hold to the unity of nature idea believe that although humans are intimately connected with nature, human society possesses its own unique characteristics which are not derived from nature. This particular way of looking at unity is used as a defence by committed unitarian holists, like Freya Matthews[4], who believe that the charge of fascism cannot stand since recognising the inherent unity of nature does not mean the recognition of, or support for, social unity.

Suppose we accept this point as valid: natural unity does not necessarily force upon human society a social unity. If we do accept it as valid, this only defends the unity of nature idea from social fascism. Claims of ecological fascism (that the unity of nature is anti-individualistic and intolerant of dissent and difference in non-human ecological settings) may still be relevant. This is the point in question in this chapter.

While numerous scientific traditions have had their romances and intrigues with natural unity, the most obvious and commonly cited example of a science that recognises the unity of nature idea is ecology. Ecology studies all sorts of things, of course, from the behaviour of single animals to the behaviour of whole landscapes; from the life histories of house flies to the trans-continental migration of birds. In all this diversity of study, however, there is, according to many philosophers of nature, a common metaphysical commitment to unity. The very essence of ecological science, it is often thought, is its recognition of natural unity:

> ecological awareness...recognises the fundamental interdependence of all phenomena.[5]

> The ecological point of view is, first of all, holistic. It focuses upon the 'all-ness' of nature.[6]

Frederic Clements and the Use of Unity as a Metaphor

One of the first intellectual endeavours which tied the unity idea to ecology was when the American prairie ecologist, Frederic Clements, developed and promoted his superorganism thesis in the early Twentieth century.[7] To Clements the prairie community of plants and animals had a structure and a physiology analogous to that of an organism. The prairie community grew like an organism, reproduced like an organism, and maintained its boundary integrity like an organism does. To modern-day environmentalist Edward Goldsmith the superorganism is what distinguishes ecology from other sciences.[8]

The use of the word 'superorganism' by Clements, and his modern-day intellectual followers like Goldsmith, might immediately strike one to believe that Clements meant that the prairies were in fact great big vast organisms but really it could mean any manner of things. The open-endedness of the metaphor allows for a variety of meanings. For instance; that the prairie has boundaries like a living organism but no physiology. Or conversely it has a physiology but no boundaries. In other words, the prairie is significantly comparable to an organism or some of the aspects of an organism but is not actually an organism in itself. The 'prairie as an organism' is mere metaphor. In the vein of Aristotle[9] we can describe it as an attempt to see similars amongst dissimilars, a quest to elucidate the unfamiliar (the properties of prairies) in terms of the familiar (the properties of an organism).

The very fact that the superorganism is a metaphor might mean to some that we should not take it too seriously. Ecological communities are only *like* organisms.

They are not put forward *as* organisms. As the modern day ecologist, Frank Golley, notes, Clements himself, "called the climax a 'complex organism' to distinguish it from the well recognised individual organism."[10]

Similar points are made by environmental philosophers as well as ecologists. For example, Laura Westra, says:

> Of course, there is no claim made here that individuals and ecosystems are analogous in all respects. It simply appears that their status as developing, changing entities containing life, renders them similar to each other in some relevant respects.[11]

This is an argument that may be raised again and again with reference to some of the points in this book since it is a work that deals with both acknowledged and unacknowledged metaphors; i.e.: 'because the metaphors are rather loose and only an attempt to explain what things are *like* rather than how things *really are*, we should restrain ourselves from investigating metaphors too intensely lest one falls into seeing too many potential reasons for their application'. 'Why investigate metaphors?' this argument would ask, when, after all, the progress of science has a way of weeding out metaphors that are wrong or inaccurate and leaving only those that truly reflect reality?[12]

If we accept that metaphor is merely a non-literal and poetic form of human expression, we are, however, still left with two problems in the use of holistic metaphors in ecology. Firstly, the metaphors that ecologists like Clements use are often not utilised as loose descriptions at all but as appeals to the literal truth. Clements' superorganisms are not just rhetorical devices that attempt to elucidate the unfamiliar in terms of the familiar but are efforts to categorise the reality of a certain circumstance. Certainly Clements would have thought of his organismic metaphors as more than explanatory tools. They were sustained attempts to positively say that something is so comparable to something else that they both belong to the same category of things.[13] The same can be said of modern day ecologists and philosophers of nature and their use of holistic metaphors. For instance, Fritjof Capra writes:

> the Earth, then, is a living system; it functions not just like an organism but actually seems to be an organism--Gaia, a living being.[14]

Similarly, Roger Lewin writes of his holist ecology colleagues:

> for Goodwin and Mitchell, the emergence of such regularities is what characterises a superorganism, rather than its being like a superorganism.[15]

Secondly, there is the question: 'how come one particular metaphor was chosen over others?' Why is the ecology of the world more like a superorganism than a machine, or river, or a whistling wind? If there are a whole lot of potential or available metaphors, why are particular ones chosen over others? These questions draw us into taking metaphors seriously.

Ecological Science and the Intellectual Development of the Superorganism

For Frederic Clements, in the early part of the Twentieth century, and for modern-day holist ecologists like Edward Goldsmith and Goodwin and Mitchell, the superorganism is supposed to be composed of individuals and species that are so interrelated and interdependent that the community or ecosystem they comprise can be regarded as a balanced, self-regulating and highly defined organic unit; a unit whose constituent species possess a certain ecological complimentarity with one another. This ecological complimentarity bestows a form of functional integration so the characteristics of stability, resilience and non-invadibility are conferred upon the ecological community as a whole. Because of all of this, an ecological community of living things some how resembles an organism. Lewin puts it this way:

> Although a precise definition is elusive, a superorganism can be thought of as a group of individual organisms whose collective behaviour leads to group level functions that resemble the behaviour of a single organism.[16]

Clements' superorganism not only exists in space; it also exists in time: a superorganism grows or evolves. This is encapsulated in Clements well-known theory of succession. Succession theory is a theory of ecological development which posits that there is progressive sequential vegetation change on any given geographical site. This change proceeds through various stages from invading pioneer species through intermediary stages to a stable, mature and relatively uniformly-arranged climax community.

Clements believed that each stage in the development of the climax community was essential. You couldn't just have a climax community, such as a stand of oak trees, pop on to a piece of ground without that ground being prepared and developed by an ensemble of species present in previous stages. Thus, for an oak forest to have developed at a particular site, a whole series of pioneer and intermediary species would have had to prepare the ground, manipulate the soil, adjust the micro-climate, and do a host of other things to create an environment suitable for the first successful invasion by an oak tree.

This process not only occurred in the making of oak forests; Clements thought that it was a near universal phenomena in all terrestrial communities from his beloved prarielands in Nebraska and Colorado to the mountain forests of Canada and the swamplands of Florida.

The succession process was not only universal but also somewhat predictable, too. Given a certain bunch of initial species, and given a certain set geographical factors (such as rainfall volume, soil type and atmospheric temperature), the sequence of succession was teleologically destined towards a single resultant community of plants (or to a very few variations of this result).

Clements was a major figure in pre-War American science, having contributed to the development of two icons of ecological theory and practice; firstly, the theory of succession just outlined; and secondly, the quadratic sampling technique.[17] With regards to the second, Clements, in his early years, at the Universities of Chicago and Nebraska around the turn of the century, was determined to turn the rather patchy intellectual profession of botanical natural history into a 'real' science; the science of ecology. According to Robert Macintosh, an ecologist turned historian, Frederic Clements:

> decried...descriptive ecology, meaning verbal descriptions of vegetation, often only accompanied by species lists. Clements wrote of such descriptive ecology that no method can 'yield results farther from the truth'...Ecology was to be, in Clements' conception, quantitative.[18]

Clements' model discipline in this endeavour to quantify ecology was physiology. By constantly measuring the physical and numerical parameters of his subject of study, i.e. the prairie communities of Midwestern U.S.A., Clements moved ecology towards what he regarded as being a real science. As he continued quantifying America's natural environment, Clements moved up the ranks of the scientific establishment and in the 1920s reached the higher echelons of American biology when he was awarded a permanent research position at the new Carnegie Institute.

By the 1940s however, the quantitative appeal of Clements' botanical studies were outshone by his growing theoretical obsession with the superorganism idea and by his inflexible commitment to succession. Although succession is still a much discussed topic today within terrestrial ecology circles, it is usually described as an ideal model and is often given only a pedagogical role. In most places where professional ecology is practiced, succession theory is thought of as being an over-generalising, deterministic view of natural communities. These problems were becoming aware to Clements' later contemporaries and very few of them actually thought succession was quite deserving of the theoretical universality that Clements attached to it.

As various chroniclers of the history of ecology have explained, some of Clements' contemporary critics, like S. A. Cain, regarded Clements' propensity to classify and categorise hundreds of different types of climax communities as self-contradictory proof that climax communities were not deterministically inclined towards a particular endpoint.[19] In contrast, Clements continued to think of both climax and succession as central concepts to the study of virtually all terrestrial ecological settings.

Although he had many supporters, Clements also had some detractors, and these grew in number as his career and life went on. It might be noted that none of this criticism and support for Clements ever fell away after his death and today we can still discern two ways of thinking within professional terrestrial ecology; that of unitarianism and that of anti-unitarianism (although these two schools of ecological thought are usually labeled using the terms of 'holism' and 'anti-holism').

The anti-unitarians are often castigated by holist thinkers as mechanistic, reductionistic, and unworthy of the title of ecologist.[20] Edward Goldsmith, for example, when writing to honor a man he perceives to be a modern day Clementsian ecologist, finds it impossible to hold off from having a stab at 'atomistic' ecology:

> I am indebted to Eugene Odum, one of the few remaining academic ecologists whose work has not been perverted to fit in the paradigm of mechanistic science.[21]

The loudness of the holist intellectual movement in ecology and environmentalism is, however, probably outmatched by the sheer number of anti-holist ecologists in the profession of ecology. Yet, although there is a lot of it, the presence of anti-holism in ecological science is generally ignored by environmentalists. Or to put it more accurately, the implications of the professional distrust of holism is ignored. Many environmental thinkers however are just about totally ignorant of professional anti-unitarianism in ecology and so go on to believe that ecology is really the only discipline in science where holism is concomitantly supported in both theory and practice.

A lot of professional ecologists would be somewhat bewildered by the continued appraisal that 'ecology equals holism' when their own professional lives revolve around working with theories that declare the subjects of their study are somewhat atomisticin nature. Such anti-unitarianism started early within professional ecological science. Scientists such as Henry Cowles, P.J. Ramensky, W.S. Cooper, Arthur Tansley, Forrest Shreve, S.A. Cain and Henry Gleason reacted, mostly gently, against Clements and his supporters' use of superorganisms.[22] The most celebrated and perhaps the most relevant of these anti-unitarians was Henry Gleason.

Henry Gleason's reactions are best recorded in a 1926 paper that has come to be regarded as a classic in the science of ecology.[23] In this paper Gleason described ecological communities not as interdependent unities but as fortuitous associations of ecologically and genetically unrelated plants brought together by chance happenings in their history and migration.

The Gleasonian view of ecological communities emphasises that the well-defined and integrated character of any particular naturally-occurring group of plants and animals is a human abstraction, and that communities are anarchic, stochastic and fluctuating. Species composition, according to Gleason, is not uniform but continuously varies over space and time, and in such a disjointed and unpredictable manner that generalisations such as succession are impossible or useless. Plant aggregates are thus boundless and bondless associations rather than hard and fast well-defined unities. Or in Gleason's own words:

> just as it is often difficult and sometimes impossible to locate satisfactorily the boundaries of an association in space, so it is frequently impossible to distinguish accurately the beginning or the end of an association in time...A community is frequently so heterogeneous as to lead observers to conflicting ideas as to its associational identity, its boundaries may be so poorly marked that they can not be located with any degree of accuracy, its origins and disappearance may be so gradual that its time boundaries can not be located; small fragments of associations with only small proportions of their normal components of species are often observed; the duration of a community may be so short that it fails to show a period of equilibrium in its structure.[24]

Another important difference between Clements' ecology and Gleason's ecology is their respective attitude to the 'unusual' in ecological situations. Whereas Clements' superorganismic theory of communities would regard fragments of vegetation that do not conform to the regional climax community as being peculiar, Gleason's outlook would see these non-conformities as typical:

> Every ecologist has seen these fragmented associations, or instances of distribution, but they are generally passed by as negligible exceptions to what is regarded as the general rule.[25]

If the superorganism and the community were Clements' main metaphors of elucidation then Gleason's main metaphor of elucidation was the 'association'. Within the framework of the association, ecological groupings of organisms may possess interdependent relationships that may even be obligatory for the individuals and species involved but the whole aggregate does not register holistic properties that confer upon it the status of a unity.

It may, at this point, be useful to concoct a definition of Gleason's association versus Clements' superorganismic unities in order to elaborate upon the difference between the two:[26]

-a 'unity' is a united entity composed of non-separable parts which act in a unified, integrated and interdependent manner, whether conscious of it or not, toward a common agenda: the maintenance of the unit-entity as a whole.

-an 'association' is a loosely-gathered group of coincidentally-arranged, separate 'unitary' organisms living together, sometimes in an interactive and interdependent way, but which act individualistically and without contribution to a common agenda.

Although Gleasonian ecological settings exist more as random and ill-defined associations rather than the bounded and integrated superorganismic unities that Clements and his followers suppose, this does not mean that Gleasonian ecological associations are never at all integrated and that the members that compose them are not highly interdependent in particular situations. Quite possibly individual members within Gleasonian associations do enter into the occasional strategic alliance in order to uphold a common interest.

It is highly improbable that any one individual exists as an isolated ecological atom in an association (although this may happen sometimes). Most individuals do interact but Gleason would say that this would be at a localised scale rather than at the whole community or ecosystem scale as advocated by Clementsian ecologists. The Gleasonian view of ecology can just as well be read as a recommendation to recognise the interactions of ecological agents but, as we shall see, Gleasonian ecology suggests that this ecological interaction is not of the type that unitarians uphold.

Divergent Types of Ecological Study

From the mid 1930s on, the debate between anti-unitarians and unitarians became significantly more complicated. The British ecologist Arthur Tansley had introduced the ecosystem concept so as to put the organic holism of Clements into a more acceptable 'scientific' framework. When Tansley put forward his ecosystem concept, he was "qualifying without disabling"[27] Clements' earlier organismic concepts.

While Tansley invented the ecosystem concept in 1935, it was not until 1941 that someone actually found one since this is when Yale limnologist Raymond Lindeman started his studies on the Cedar Creek Bog ecosystem in Minnesota. Instead of studying the incidental organisms in this bog, Lindeman studied, and more significantly, he measured, the bog's energy characteristics.[28] By doing this--by asserting that energy flowed throughout the ecosystem--Lindeman came to realise that everything was interconnected and united within the bog because of this flow. This uncompromising appeal to quantitative energy measurement and tracking distinguished Lindeman from his contemporaries who were unready to apply Tansley's ecosystem ideas to their own field studies. Hagen confirms this when he says:

> what most clearly distinguished Lindeman from older ecologists was his use of the ecosystem concept. With this concept he was able to synthesise elements taken from traditional studies in fisheries biology and limnology, the newer biogeochemical approach to studying lakes,...terrestrial ecology and traditional plant ecology.[29]

In the 1950s and 1960s:

> most studies essentially repeated Lindeman's study of Cedar Bog lake, in that they aggregated the diverse fauna and flora into a small set of trophic groups, determined the flows of energy and materials between groups, calculated ratios of inputs and outputs.[30]

Lindeman's observation and quantification of energy flow served to revitalise superorganismic ideas in ecology since there was now something organismic out there in the real world to observe and quantify: ecosystems.

As indicated previously, there have been two differing viewpoints within the science of ecology that roughly correspond to an adherence to either a superorganismic outlook or to an association outlook. Since Lindeman's time these two differing viewpoints have paralleled the different approaches between two developing breeds of professional ecology; namely 'ecosystem ecology' (the ecology of superorganismic ecosystems) and 'community ecology' (the ecology of individuals making up an ecological community). These divergent fields of ecological study had different attitudes to the superorganism idea and they also developed divergent attitudes to Clements' succession theory:

> while neo-individualist theories tend to view succession as a consequence of the plants that dominate a system the ecosystem theorist might view the successional change as being manifested in the balancing of production and respiration or of equilibrium of input and output of major nutrients.[31]

One of the interesting points to note when discussing this divergence in ecological science is that community ecologists have never really believed in the concept of the community, it was just a convenient spatial abstraction to them, whilst ecosystems ecologists believed in both the community and ecosystem concepts as being mirrors of an independent ecological reality. For ecosystem ecologists 'the ecosystem is the more fundamental unit'[32] of ecology but they are still inclined to invoke the community concept as defining the biological component of that ecosystem.

For community ecologists, on the other hand, it is often the regarded that the various organismic members of a community, and their immediate relationships, are the fundamental units of ecology. Given their chaotic, ill-defined state, however, communities themselves are not really there in any physical sense.

Whereas community ecologists usually hail from traditional zoology and botany schools (or ecology departments that evolved from such schools) ecosystem ecologists are just as likely to be physical or mathematical scientists as experienced biologists. Because they are not content to set about describing the enormous variability in the ecological world by the pragmatic and fragmentary methods of botany and zoology, ecosystem ecologists have attempted to harden the discipline of ecology into a 'real' science; one replete with universal scientific laws.

Peter List argues[33] that ecologists have never arrived at a scientific law as one finds in the physical sciences; ecological laws are much too flimsy or specific. This criticism--which emerges again and again from within and without the ranks of ecology--has prompted the (now-aging) project of transferring physical science techniques (especially those of geochemistry, process engineering, network analysis and cybernetics) into the study of ecological subjects.

Although the critique of ecology's inability to formulate basic laws continues and although ecosystem ecologists continue to try and physicalise and quantify their subjects of study to arrive at those laws, community ecologists generally ignore the call to physicalize their science and are content to just go merrily along on their way describing the actions and interactions of small collections of living things rather than trying to find universal theories.

This difference in ecological approach started about a decade after Lindeman's first Cedar Bog study, since:

> in the 1950s Hutchinson and Wollack...identified a dichotomy in ecology which was increasingly evident as ecosystem ecology developed. One method of ecology, they said, concentrated on the 'biosociological', based on individual species and their relations. The other isolated a space and studied the transference of matter and/or energy across the boundaries of space.[34]

While the superorganismic concept was suffering intellectual suspicion during Clements' later years in the 1930s, it was, after Tansley and then Lindeman, to be reconstituted and reinvigourated through ecosystem ecology. As Richardson argues[35] the ecosystem approach is, at least in part "a child of the organismic concept--a child that in turn has nurtured its embattled parent."

Systems Ecology and The Unity of Nature

According to Fritjof Capra:

> A system has come to mean an integrated whole whose essential properties arise from the relationships between its parts, and 'systems thinking', the understanding of a phenomenon within the context of a larger whole[36]

At the extreme end of ecosystem science is what is called 'systems ecology'. Systems ecology is often perceived as being distinct from ecosystem ecology (the latter being a biological discipline which makes extensive use of physical science methods to analyse ecological situations whereas the former defines biological and ecological activity purely by the operations of physical processes discovered by such methods) but the division between the two is somewhat fuzzy. They exist side by side with regards to having many common analytical techniques and similar conceptual frameworks.[37] Thus any criticism directed specifically at systems ecology or ecosystems ecology must surely fall upon the both of them.

Within the metaphysics of systems ecology, all living things on the planet are united and unified due to the physical transfer of materials and energy between the system constituents. The cycling of 'nutrients' (a term that embodies both matter and energy) is a key focus of study in the unity approach, as Tyler Volk makes clear in his short exposition of the work of two particular systems ecologists:

> Tim Allen and Thomas Hoekstra see nutrient cycles as the very markers designating the existence of ecosystems.[38]

To Frank Golley the predilection within systems/ecosystems ecology was for transferring physical ideas into the biological domain. This approach:

> tended to de-emphasise the significance of biological differences. Species and individuals were represented as mass, energy or chemical elements...Although the advantages were many, the disadvantage was that most of the biological reality encompassed in the species was lost.[39]

"In the ecosystem model", carries on Golley "species acted abstractly like robots".

When talking about the whole planet, the term biosphere sometimes rears its head in systems talk. The biosphere is the largest ecosystem and it comprises all living things on the planet plus the geochemical systems that they are a part of. By systematising the whole planet, all the different and separate members of that biosphere become functionalised, i.e.: placed into categories and awarded a role. Rather than existing as distinct living beings, individual organisms (and the local collections that they comprise) are turned into components of a system. Each component is then judged by its contribution to the continued cycling or transfer of matter and energy within the system. Often the exact species identity is of no importance in this process of systematisation. Such applies, for instance, to the work of the Eugene and Howard Odum, two brothers within 1950s and 1960s ecology who are often held to be the flagbaring pioneers of systems ecology:

> When the Odums had studied the metabolism of the reef at Eniwetok Atoll, they were not concerned with individual species. Indeed, at the time they were unable to identify them.[40]

During the course of their research and analysis, the conceptual framework used by systems ecologists rapidly transforms individual living beings and collections of living beings into typological categories that designate their particular role in the ecosystem.

For, instance a birchwood tree becomes a carbon-fixing autotroph which transforms so many photons of light into carbon over so many years, a beaver becomes a heterotrophic carbon consumer releasing *x* amount of calories into the system cycle per unit time, and a bracket fungus becomes a decomposing component working at so and so rate of efficiency as ascribed by its functional status. It matters little whether the plant is a birchwood, an exotic alien plant, a vat of industrially-grown seaweed or a great amorphous blob of plant cells, since its function as a photosynthetic carbon producer is all that is cared about.

This link between unity and function is also emphasised by some environmental philosophers. For instance, Laura Westra shows a predilection toward unitarian functionalism by saying "integrity conveys the idea of wholeness and unbroken functioning."[41]

Given the systems stance, it might be suggested that a change in the biotic construction of an ecological setting (such as species composition) is unimportant if the matter and energy flow systems are maintained. For example, American ecosystem enthusiast Robert O'Neill is cited as taking such an extreme view as this for he is said to maintain that ecosystems retain their character even when the individual or species composition changes.[42]

Systems ecologist, Pat Klinger, also tends to do this. Klinger is quoted by Roger Lewin as saying:

> that is because, structurally they are all the same. Yes, the species may be different in different parts of the world, but often the genera are the same. And in terms of physical form--the shape of the mosses, the sedges and other bog species--they are the same the world over.[43]

This fixation of genera over species shows the primacy of the general over the specific within systems ecology. When Lindeman was unearthing his handful of mud from a Minnesota Cedar Bog he might as well have been unearthing it from a Congolese bog; so common are their attributes, according to Lewin and Klinger.

A similar tendency towards the primacy of generalisations can be seen in the following passage by Edward Goldsmith (which tries to prove the unitary nature of ecosystems and communities):

> A number of experiments have been carried out to determine whether ecosystems display resilience stability and, if so, whether this can be attributed to their own efforts--and hence whether they are cybernetic or self-regulating systems, capable of maintaining homeostasis. The best known of such experiments are those conducted by Simberloff and E.O.L. Wilson. They removed all the fauna from several small mangrove islets and then closely watched the way they were re-colonised by terrestrial arthropods. They established that although the islets were eventually populated by very different species from the original ones, the total number of species was very much the same.[44]

Goldsmith is a very strong advocate for unity and the superorganism idea but his preoccupation with unity and with finding generalisation means that he ignores the very differences that exist within ecological settings. Species differences in any particular comparison between two different ecological settings are cast aside by Goldsmith in order for him to make a statement about the ecological reality of unity.

Reiterating Goldsmith's preoccupation with overall structural similarities (instead of specific differences), is Klinger:

> Klinger strongly believes that the robust dynamics of succession towards peatland displays the characteristics of a complex system, partly because they promote their own formation but also because they are so similar in fundamental structure.[45]

The fact that Klinger sees all peatbogs as exhibiting just a few community types shows his assumptions towards climax ideas and Clementsian superorganicism since Clements' succession theory also presupposed a few common types of climax community. Klinger admits as much to Lewin, who says of his work: "Klinger's peat bogs figure in the current revival of the idea of the superorganism."[46]

The Association Perspective: Gleason's Ecology of Individuals.

In contrast to the unity perspective of superorganicism one of the important ideas to emerge out of the association way of looking at ecological communities is that when the individual or species composition of an ecological community changes, then the whole community can be considered to have changed. This is not merely a statement of the value of individual organisms, it can also be considered a description of ecological reality, since:

> as far as is known, each species is uniquely different from each other species in at least some respects. This means each species must respond uniquely to the ecological situation in which it occurs, compared with other species exposed to the same set of conditions. In fact...genotypic variation within species and even within individuals will produce some ecological difference between populations of the same species.[47]

This statement, from a community ecologist, concurs with another, from an environmental philosopher:

> To speak of human uniqueness is quite acceptable as long as we are prepared to accept that chimpanzees, dolphins, bees and humming-birds are also unique.[48]

Such an overwhelming quantity of uniqueness within and between species would surely suggest that any competent ecological science would acknowledge the sheer complicated texture of the ecological world. Schulze and Zwolfe, for instance, maintain that the reason for difficulties in formulating universal laws in theoretical ecology is the sheer diversity of the behaviour, interactions and adaptations of individual species.[49] If the diversity between and within species is so great then far from retaining its character in the face of a change to its species composition, as Robert O'Neill would assert, species composition is an important defining feature of ecological communities.

It may also be the case that the personal history of an individual (and the localised context in which it is embedded) might suggest that uniqueness is an individual property and not just a species property. For instance, if we could analyse twin plants who had developed in different environments we might find that each plant would react to a common environmental event (say a decrease in predation rate or an increase in rainfall) in a different way. This effect would not come from genetic heritage but from the lived experience of the plants.[50] From this point of view it seems fair to say that the character of any one ecological community must be peculiar to that particular community not only because it has a unique assemblage of unique species but also because it has a unique assemblage of unique individuals.

While the composite members of an ecological community all possess their own unique character, for nearly all individuals this character is not independent of their environmental context. This is to say that there are interactions within the abstract community that have the capacity to influence the nature of the individuals and species within it. This is done through the medium of emergent properties. Emergent properties such as soil fertility, nutrient availability, predation rates, niche patterns and resource distribution commonly affect the character of individuals within a community.[51]

Herein arises an apparent paradox in my interpretation of Gleason's associations for while emphasising an appreciation of atomism in ecology (so that large community or ecosystem wholes are counted as abstracted or constructed entities) there is, nevertheless, a willingness to attribute super-structural feedback upon individuals via a holistic and unitarian-like concept such as 'emergent properties'. However, it is my view that an individualistic view of ecological communities can co-exist with the concept of emergent properties.

Although ecological communities are justly regarded as abstractions of the lives of individuals and their interactions, ecological communities do possess properties that any individual cannot exhibit on its own. This is to say that emergent properties are not dependent upon these communities being hard and fast self-organising and unitary entities. It must also be emphasised, though, and this is a crucial point, that emergent properties are just as heterogeneous, variable, self-contradictory, transient, and evasive of generalisations as the associations from which they emerge; such that patterns of feedback can only be ascribed very tentatively to any ecological situation.

Systems Ecology: A Reductionist Type of Holism?

Systems ecology claims to be a holistic approach to ecology, since it looks at the total array of material cycles and energy flows going on within the world. However, any ecological approach based purely upon the study of matter and energy is an approach rooted to reductionism since ecological actors and processes are reduced to physical entities only. Or to put this another way; much of systems ecology is not ecological, since the phenomena being explained "can be evaluated by physical and chemical methods, with total disregard for ecological ones."[52]

By reducing living entities to physical entities, by producing physical theories of the ecological world, systems ecologists can be said to be suffering from what is often called 'physics envy'. Many scientists would not see much of a problem with such a physical approach to the living world, since it paves the way to understand the processes going on in terms of well-established physical laws. But for social scientists it might seem less benign since it is akin to studying a human community by examining the flow of electricity and food into it and the heat waste and material refuse out of it. This might be a useful exercise but not one that many social scientists would think of as having described the total character of the community.

While the physical approach of systems ecology might be an adequate explanation for the integrated nature of a unitary individual organism--where the transfer of matter and energy proceed without a hitch according to physical laws--it is not an adequate explanation for the activities of an association of distinct individuals; where the phenomena involved with getting into or out of a position that enables the transfer of matter and energy are more important in determining ecological structures and biotic relationships than the mere transfer of matter and energy itself. Such phenomena (like predator-prey interactions, competition, mutualism and parasitism) are not reducible to physical laws. As a demonstration of this consider a forest community. The ecological structures and natural relationships within a stand of forest trees are not determined by the physical connection that photosynthesis might be said to effect between a tree and the physical components of its environment (light, water, air, etc.) but are determined by a great range of factors which involve the tree attempting to effect such a connection, including, for example: subsurface water competition, above-ground competition for light, herbivory evasion, symbiotic partner availability, parasite avoidance, and chance confrontations with ecological disturbance.[53]

None of these processes act according to known laws of physics and so they cannot be incorporated as subsystems into a total systems model of the forest, yet they are far more important when describing the structure of the forest stand than the physics and chemistry of photosynthesis.

We might also find when adhering to the physical approach of systems ecology, many relationships that are crucial in determining the emergent (though transient) structures of the Earth's ecological communities are ignored because they do not significantly contribute to the flow of matter and energy. The most obvious examples are those relating to sexual interaction and reproduction such as pollination, seed dispersal, courtship display and off-spring care.[54]

Ecological Unity and Hierarchy

Nature's unity is often organised into distinct layers by unitarians. The way they do this is by the use of hierarchical levels (sometimes called 'levels of organisation'). Under this schema the various units that make up the whole entire unity of the natural world are distributed hierarchically onto levels.

According to Capra:

> an outstanding property of all life is the tendency to form multileveled structures of systems within systems. Each of these forms a whole with respect to its parts while at the same time being a part of a larger whole. Thus, cells combine to form tissues, tissues to form organs, and organs to form organisms. These in turn exist within social systems and ecosystems.[55]

C.D. Rollo explains hierarchical levels this way:

> A hierarchy is a system of organisation in which different levels can be distinguished or where lower levels are sequentially nested within levels above.[56]

When Rollo says this, it is easy to see that he could find an ally in Edward Goldsmith who lets us know that: "the biosphere is a hierarchical organisation of natural systems."[57] The systems ecologist, George Van Dyne would described this hierarchy of increasingly complex levels by pointing out its structure as being:

> cell<tissue<organ<organism<population<community<ecosystem[58]

The ever-present nature of hierarchies in unity theory is of much importance and will be focused on more than once in this book. My purpose here, however, is just to draw the readers attention to the fact that unity and hierarchy may be connected concepts.

The Order of Things

Environmentalists that operate under the ontological reality of the natural unity also subscribe to metaphysical companion concepts that prop up and support unitarianism. Along with unity, nature is held to exist as a naturally stable, self-regulating, orderly and harmonious balance. Thus stability, harmony, balance and self-regulation all exist with and within the unity of nature idea. It should be noted, however, that these ideas have been vigorously attacked by various non-systems thinkers within the science of ecology. In this section these adjunct concepts of unitarianism are examined one by one and contrasted against the Gleasonian alternatives. In explicating these alternatives, I am not just putting forward the case for the comparable heuristic value of the Gleasonian alternative, I am also stating that the first pair of all the binaries has been a privileged focus within Clementsian, ecosystemic, and unitarian research.

a) stability versus change:

According to Edward Goldsmith "stability rather than change is the basic feature of the living world."[59] The idea that natural communities are stable entities is exemplified in the popular environmental expression: 'the balance of nature'. Although the balance of nature concept has permeated the study of the natural world since antiquity[60] it has become scientised through Clementsian plant ecology and through ecosystem ecology; both of which perceive that balance and stability are common characteristics of ecological phenomena of many types and at different levels.

While stability and balance may appear to be prevalent phenomena at various scales in nature; stability and balance as pervasive explanatory devices are inadequate when confronted by the many instances of instability and imbalance recorded by various observers of nature.[61] Even dynamic stability (whereby periodic and episodic cycles of change proceed within certain parameters) can not necessarily be regarded as a fundamental characteristic of biotic collections. Indeed, according to Gleasonian ecologists, stability may only ever be an ephemeral phenomena of but a few communities.[62]

John Wiens[63] points out that many ecologists presume that stability and balance are directly relevant to the study of any particular ecological community and because of this presumption they tend to use methodological techniques that automatically find stability. Connell and Sousa in turn point out that:

> natural perturbations are often so frequent that there is not enough time for a community to achieve a stable equilibrium state.[64]

Another point of significance was made by P.W. Frank[65] in the 1960s who indicated that humans tend to infer a state of balance and stability within ecological communities because the communities are composed of long-lived individuals whose life-span is many times that of a human. A forest or a coral reef only seem stable because humans live such short lives.

It is important to recognise that the concepts of ecological stability and/or ecological equilibrium often mean different things to systems ecologists and community ecologists. This involves a perceptual difference in scale as well as a difference in what Connell and Sousa call the 'characteristics of interest'. An ecosystem might be considered by a systems ecologist as being in energetic or nutrient stability but its species composition and biotic structure may be changing all around.

With regards to scale, it might be asserted that stability is not a standard feature at the population or community level of organisation but becomes more prevalent as one observes at greater and greater scales. In other words; those that do not see stability and balance are not thinking big enough, either spatially (to take into account the whole biosphere) or temporally (to take into account evolutionary history). However, as the examples in the notes show, stability at even these grand scales is not assured.

b) homeostasis versus unregulated chaos

Not only is nature described as being in a stable and balanced unity; unitarian environmentalists and systems ecologists hold that the stability of that unity is a self-regulated stability. The process of homeostasis, which biologists have for along time pronounced to exist in individual organisms, is taken from this narrow biological level and applied to collections of individual organisms. Of this homeostasis Capra says:

> The flexibility of an ecosystem is a consequence of its multiple feedback loops, which tend to bring the system back to into balance whenever there is a deviation from the norm due to environmental conditions.[66]

According to the Deep Ecologist, M.O. Hallman[67], ecological science emphatically demonstrates that "the entire biosphere is composed of delicate homeostatic mechanisms which go to make up the balance of nature". The 'goodness' of such self-regulatory homeostasis is encapsulated in the well-known Leopoldian Land Ethic; which prescribes that a thing is right when it tends to preserve the integrity, stability and beauty of a natural community and is wrong when it tends to otherwise.[68]

Doubts about the existence of homeostasis beyond the organismal sphere, however, are rather strong within the science of ecology. Colin Burrows for example, states:

> Ecosystems are not really homeostatic because the plants and animals of which they are composed are not capable of the degree of communication, cognition and organisational foresight which would be required to achieve this.[69]

Even if inter-organismal communication, cognition and foresight were achievable in a natural collection of disparate organisms these factors could not surmount the incommensurable differences of interest that exist between different species and individuals.

Whether or not balance, stability or equilibrium exist as fundamental features in an ecological collection, the mere presence of stability is not, in itself, adequate evidence for the operation of superorganismic homeostasis. As biologist George Williams points out; just because the input of a particular chemical element is equivalent to the output of the same element it does not mean that that particular system is maintaining the balance.[70] In contrast to homeostatic unity, the Gleasonian acknowledgement of biotic disunity and contradictory feedback patterns in ecological communities would encourage a view that the ecological interactions we see in ecological communities are unregulated and unrestrained. Homeostatic feedback cycles may exist but they are spatially and temporally transient and may have little or no relevance to the compositional character of communities. Indeed the ecologists Wu and Loucks admit that:

> direct evidence that ecological systems are inherently in equilibrium, however, is still lacking. Indeed, individual organisms may be the only systems within which homeostatic mechanisms have been demonstrated to operate.[71]

c) deterministic maturation versus non-determinism

Systems unitarianism, through its adoption of Clementsian ecological principles, also has a strongly deterministic stance. Clementsian ecology and ecosystem ecology both affirm the existence of ecosystemic or community maturity via the concept of the ecological climax. The ecological climax is perceived to be the final teleological end-state within a succession of natural communities.

However, such ecosystemic maturity seems, to Gleason and the Gleasonians, more likely to be a human abstraction. Maturity, like balance, stability and homeostasis, is only ever a phenomena of unities and not associations. The Clementsian theory of succession states that a wonderfully ordered and sequential development of an ecological system/community occurs as it ages (characterised by increases in biomass, species diversity, community stability, community self-regulation, biotic complexity and productivity) as it proceeds from pioneer stages through intermediary stages to a mature climax community.

However, that these developmental tendencies actually occur in a majority of ecological settings is debatable. Correlations between aging and increases in biomass, diversity, stability, production etc. are often non-existent in many (maybe most) natural communities. For many contemporary ecologists the succession theory in one form or another still stands but others feel that it has more or less been slain. For them, the term climax must always risk being a misnomer when applied to particular ecological situations. There is no succession, only change.

Summary: The Unity of Nature as Ecological Fascism

From an intellectual perspective, it seems that there is reason to regard the unity of nature concept with suspicion. We can no longer claim that the science of ecology indisputably supports the contention that the living world is united. However, from an ethical perspective there is also a case to place doubt upon the unity of nature story, at least as it is narrated by holist scientists like Clements and his intellectual off-spring. Within their hands the unity of nature idea seems to be fashioned in a decidedly fascist way. This fascism comes about not because of what the unity of nature might do to the human world but what it does to the non-human world; creating abstract wholes and functions while disregarding difference and abnormality, appealing to the primacy of the general over the uniqueness of the specific.

To derive values from the worldview of unity may be to stumble at the starting blocks of one's environmental evaluation. Neither the united superorganism nor the united ecosystem can direct us towards learning about the very things that most ecologists believe make up the environment, namely: the unique individual members of that environment and the various, indistinct, collections that they make up. All in all, systems ecology seems to de-ecologize ecology, turning it into a science of physical, rather than ecological, processes. If environmentalists choose to adopt systems ecology as their primary metaphysical stance then they may be adopting a worldview which is inimical to valuing the parts of the natural environment that they are usually so committed to protecting.

In order to reconfigure this lack of concern for the specific and the unique within unitarian and systems ecology, and in order to develop the unity of concept into a worldview that can appreciate instability and imbalance, it may be wise for environmentalists to take note of Clements' intellectual adversary, Henry Gleason. Unlike Clements and his systems ecology heirs, Gleason's association ecology appreciates the shear complicated variety which confronts the ecological observer.

This idea, that Gleasonian ecology, rather than Clementsian ecology, could serve as the ecological site of an environmental narrative is fully explored in Chapter Eight. However, before then, other social and philosophical problems that follow unity around are examined.

Notes to Chapter 2 (Falling Into Wholes: Ecological Fascism)

1. D. Robertson, *Penguin Dictionary of Politics* (London: Penguin, 1993 edn) p183.
2. J. Macy, "Toward a Healing of Self and World", in C. Merchant, ed, *Key Concepts in Critical Theory: Ecology* (NJ: Humanities Press, 1994), p.297.
3. D.E. Marietta, "Environmental Holism and Individuals", in S.J. Armstrong and R.G. Botzler, eds, *Environmental Ethics: Divergence and Convergence* (N.Y.: McGraw-Hill, 1993) p.407.
4. F. Matthews, *The Ecological Self* (London: Routledge, 1991).
5. F. Capra (1994) "Systems Theory and the New Paradigm", in C. Merchant, ed, *Key Concepts Critical Theory: Ecology* (NJ: Humanities Press, 1994) p. 335
6. E. Partridge, "Nature as a Moral Resource", *Environmental Ethics*, 6, (1984) :106.
7. See F.E. Clements, *Plant Succession: An Analysis of the Development of Vegetation* (Washington, D.C.: Carnegie Institution, 1916).
8. E. Goldsmith, *The Way: An Ecological Worldview* (NY: Shambhala, 1993), p15.
9. See Aristotle's *Poetics*, in, for example, Aristotle, *The Basic Works of Aristotle* (NY: NY Books, 1941). Don Miller points out that Aristotle's is still the dominant Western configuration of metaphor and he points us towards the *Oxford English Dictionary* as an example of how this configuration is defined. (See D. Miller, Metaphor and Culture, *Paper Presented to the 1983 Culture Seminar*, University of Melbourne, Australia). The *OED* states that a metaphor is: 1) "the application of a name or a description to something to which it is not literally applicable" or 2) "an instance of this" (See D. Thompson, ed, *The Oxford Dictionary of Current English* (Oxford: Oxford University Press, 1996, p55.)
10. F.B. Golley, *A History of the Ecosystem Concept in Ecology* (New Haven: Yale University Press, 1993) p.24.
11. L. Westra. *An Environmental Proposal for Ethics* (Baltimore: Rowman and Littlefield, 1994) p. 43.
12. This is a simple expression of one of the two main contemporary philosophical conceptualisations of metaphor (as noted, for example, in Miller, Metaphor and Culture, and D. Miller, "Metaphor, Thinking and Thought", *ETC.: A Review of General Semantics*, 39, 1982, :134-50). Derived from Aristotle's *Poetics* this view of metaphor holds that metaphor is a primitive, imperfect, pre-scientific, rhetorical, polysemic, artistic, figurative, non-representational, poetic tool in human speech, behaviour and activities that can be contrasted with rational, scientific, literal, logical, natural, linear, monosemic, referential, representational language and thought upon which our proper attention should be focused. From this perspective metaphor, while clever and poetic, is also a danger to scientific discourse since, although it may possibly make things easier to comprehend, too often it fogs over what is literally true with various kinds of rhetorical wizardry. The other main contemporary philosophical conceptualisation of metaphor declares that metaphor is not just an inferior poetic form of expression subservient to representational language and thought but that metaphor is the only available form of expression. From this point of view--which Miller promotes and with

which I have much sympathy--representational language and thought is but a part of metaphor; a part that, according to Miller, denies its own metaphoricity in order to increase the depth of its knowledge claims. Miller's ideas upon metaphor feed into those examined later in this work: see Chapters Five, Six, Seven and Eight. Miller himself acknowledges the influence of a diverse variety of scholars of metaphor in his work; these include: Frederick Nietzsche, Max Black, Jacques Derrida, Paul Ricouer, George Lakoff and M. Johnson. For works on metaphor by these scholars see: F. Nietzsche, "Philosophy and Truth", in D. Breazeal, ed, *Selections from Nietzsche's Notebooks of the Early 1870s* (NJ: Humanities Press, 1979); M. Black, "More About Metaphor", in A. Ortony, ed, *Metaphor and Thought* (Cambridge: Cambridge University Press, 1979); J. Derrida, "White Mythology", *New Literary History*, 6, 1 (1974), 4-74; P. Ricouer, *The Rule of Metaphor*, (Toronto: University of Toronto Press, 1977); G. Lakoff and M. Johnson, *Metaphors We Live By* (Chicago: University of Chicago Press, 1980).

13. That Clements' use of holistic ideas are sustained attempts to codify the superorganism into the realm of the literal, rather than a passing rhetorical strategy, can be realised if we acknowledge the length of time that Clements stuck with it. From Clements' work in 1916 (Clements, *Plant Succession: An Analysis of the Development of Vegetation*) through to Clements' publications in the 1930s (for example, F.E. Clements, "Nature and Structure of the Climax, *Journal of Ecology*, 24, 1936, 252-284) we can note that he had invested much of his professional life championing the literal truth of the climax community as a real superorganism.
14. F. Capra, *The Turning Point* (NY: Simon & Schuster, 1982) p308.
15. R. Lewin, "All for One, One for All" *New Scientist*, 152 (14 Dec 1996), p29.
16. *ibid*, p.31.
17. See F.E. Clements, *Research Methods in Ecology* (Lincoln: University Printing Co, 1905).
18. R.P. McIntosh, *The Background to Ecology* (Cambridge: Cambridge University Press, 1985) p.77.
19. See: R. Tobey, *Saving the Prairies: The Life Cycle of the Founding School of American Ecology, 1895-1955* (Berkeley: University of California Press, 1981); R.P. McIntosh, *The Background to Ecology*; J. Hagen, *An Entangled Bank: The Origins of the Ecosystem Concept* (New Brunswick: Rutgers University Press, 1992);, *A History of the Ecosystem Concept in Ecology*.
20. The world, the holists often believe, is made a poorer place by the presence of such 'atomistic' ecologists in it. Goldsmith and Rowe, for example, are adherents of this position. See: Goldsmith, *The Way*; S. Rowe, "From Reductionism to Holism in Ecology and Deep" *The Ecologist* , 27, 4 (1997) 147-151.
21. Goldsmith, *The Way*, p.xix
22. See R.P. McIntosh, *The Background to Ecology*.
23. H.A. Gleason, "The Individualistic Concept of the Plant Association" *Bulletin of the Torrey Botanical Club*, 53, (1926):1-20. Gleason's 1926 paper is reprinted in the following anthology of classic ecological papers: L.A. Real and J.A. Brown, *Foundations of Ecology: Classic Papers with Commentaries* (Chicago: University of Chicago Press, 1991).

24. Gleason, "The Individualistic Concept of the Plant Association", p.13.
25. *ibid*, p.12.
26. While acknowledging that this may be dangerous since singular definitions often do not do justice to the myriad uses of the terms that they attempt to define.
27. Hagen, *An Entangled Bank*, p.80.
28. See R.L. Lindeman, "The Developmental History of Cedar Creek Bog, Minnesota", *American Midland Naturalist*, 25 (1941):101-112; R.L. Lindeman, "The Trophic-Dynamic Aspect of Ecology", *Ecology*, 23 (1942):399-418.
29. Hagen, *An Entangled Bank*, p.97.
30. Golley, *A History of the Ecosystem Concept in Ecology*, p.72.
31. H.H. Shugart, *A Theory of Forest Dynamics: The Ecological Implications of Forest Succession Models* (NY: Springer-Verlag, 1984) p.19.
32. Lindeman, "The Trophic-Dynamic Aspect of Ecology".
33. P. List, *Radical Environmentalism: Philosophy and Tactics* (CA: Wadsworth, 1994).
34. McIntosh, *The Background to Ecology*, p.133.
35. Or so it is argued in: J. Richardson, "The Organismic Community: Resilience of an Embattled Concept", *BioScience*, 30, (1980) :466.
36. Capra, "Systems Theory and the New Paradigm", p.27.
37. For instance, they share a penchant for mathematisation, quantification, the tracking of material exchange within ecological settings and the use of models of cycles and flows.
38. See T. Volk, *Gaia's Body: Toward a Physiology of the Earth* (NY: Copernicus, 1998) p.53. Allen and Hoekstra's unification of ecology via the use of nutrient cycling studies (and ecological hierarchy) is elaborated in: T.F.H. Allen and T.W. Hoekstra, *Toward a Unified Ecology* (NY: Columbia University Press, 1992).
39. Golley, *A History of the Ecosystem Concept in Ecology*, p.80.
40. Hagen, *An Entangled Bank*, p.139.
41. Westra, *An Environmental Proposal for Ethics*, p.xi.
42. C.J. Burrows, *Processes of Vegetation Change*, (London: Unwin-Hyman, 1990) p.426
43. Lewin, "All for One, One for All", p.31.
44. Goldsmith, *The Way*, p.131.
45. Lewin, "All for One, One for All", p.31.
46. *ibid*, p.28.
47. Burrows, *Processes of Vegetation Change*, p.82.
48. S. Horigan, *Nature and Culture in Western Discourses* (London: Routledge, 1988) p.105.
49. E.D. Schulze and H. Zwolfer, eds, *Potentials and Limitations of Ecosystem Analysis* (Berlin: Springer-Verlag, 1987).
50. Evidence which might support such a conclusion can be offered in the case of plants which produce toxins in reaction to the specific predation episodes they experience. For instance, some plants react to such predation according to the exact nature of previous predation episodes. If such a plant had a twin who was predated upon differently during these previous episodes, then each individual may react to identical ecological situations in different ways given their past variations in lived history. For those who might put great trust in the stories of practicing ecologists it might also be noted here that one of the strongest pieces of evidence which casts ecological communities as being decidedly more Gleasonian than Clementsian (i.e.: as being individualistic rather than united)

comes from observations involving the colonization of new areas. Examples of such new areas include the land created by new volcanic islands, the rocky areas beneath receding glaciers, or the barren ground of a felled forest. In areas such as these, community ecologists have sought to find out whether successive invading individuals of particular species were either influenced by--or cast influence upon--individuals of another species that either preceded them or followed them. What has often been found is that the successional processes did not depend on the presence or absence of particular species (see Burrows, *Processes of Vegetation Change*). This means that when a pioneer species invades a new site, it does not seem to prepare or facilitate further invasion by other species, as is often assumed in Clementsian succession theory, since even if the pioneer species did not arrive, a second-commer species would arrive anyway and establish itself. Thus, pioneer species weren't pioneering the land for others, they were merely pioneering it for themselves. Pioneer, in this context, just means 'the first there', rather than the 'pathfinder' or the 'trail-blazer'. The secondary invader species did not have their prospects either enhanced or diminished by the pioneers, they just took a little bit longer to grow (because of their own peculiar morphology or usually) or they took a little longer to get there (because their mode of transportation was slower). Similarly any species that arrived after the second-commers were not affected, either positively or negatively, by the presence of the pioneers or second-commers, they just took a bit longer to arrive, germinate, establish and grow. The first-commers, in turn were not affected by the late-commers, since they had either expired through the possession of a far shorter life span, or they continued to exist within a community that included late-commers. From this point of view, no real interaction is taking place to create an integrated climax community. In fact, succession, from this viewpoint, is just the gradual establishment of a myriad of species, the slowest growing and longest lasting of which just happen to be of interest to humans since they are either the most visible, the most economic important, or the most spatially-dominant.

51. The concept of emergent properties finds it way into ecological discourse via such treatments as: R.C. Richmond *et al* "A Search for Emergent Competitive Phenomena: The Dynamics of Multispecies Drosophila Systems", *Ecology*, 56, (1975) :709-714; T. Fenchel and F.B. Christensen, *Theories of Biological Communities* (N.Y.: Springer-Verlag, 1976); G.W. Salt, "A Comment on the Use of the Term 'Emergent Properties'", *American Naturalist*, 113, (1979), p.145-148; M.M. Edson *et al*, "Emergent Properties and Ecological Research" *American Naturalist*, 118, (1981):593; J.A. Drake, "Communities as Assembled Structures: Do Rules Govern Pattern?" *Trends in Ecology and Evolution*, 5, (1990):159-64. In most of these works, emergent properties are regarded as the features of a community or ecosystem which amount to more than the sum of the effects created by its constituent member species/organisms.
52. P. Trojan, *Ecosystem Homeostasis* (Warsaw: Junk, 1984)
53. On the importance of these aspects in the creation of particular ecological settings, the following publications may be reviewed:
    -On 'competition' see Richmond *et al* "A Search for Emergent Competitive Phenomena: The Dynamics of Multispecies Drosophila Systems", D. Tilman, *Dynamics*

Carney, "On Competition and the Integration of Population, Community and Ecosystem Studies", *Functional Ecology*, 3 (1989):637-641.

-On 'herbivory' see M.J. Crawley, *Herbivory: The Dynamics of Animal-Plant Interactions* (Oxford: Blackwell, 1983).

-On 'parasitism' see R. May and R.M. Anderson, "Population Biology of Infectious Diseases: Part Two", *Nature*, 280, (1979):455-461.

-On 'disturbance' see J. Heinselman and I. Wright, "The Ecological Role of Fire in Natural Conifer Forests of Western and Northern North America', *Quaternary Research*, 3, (1973):317-513; P.S. White "Pattern, Process and Natural Disturbance in Vegetation" *Botanical Review*, 45, (1979):229-299; S.T.A. Pickett and P.S. White, eds, *The Ecology of Natural Disturbance and Patch Dynamics* (Orlando: Academic Press, 1985).

54. In this regard, social scientists might note a parallel between systems ecology and orthodox Liberalist and Marxist sociology. All of these scholarly traditions tend to give little or no cognizance to the sexual aspects involved in the subjects of their study. The exclusion of sexuality thus presents a deficient interpretation of the localised lives of most individuals.
55. Capra, "Systems Theory and the New Paradigm" p.28.
56. C.D. Rollo, *Phenotypes: Their Epigenetics, Ecology and Evolution* (London: Chapman and Hall, 1995).
57. Goldsmith, *The Way:* chapter 38 title.
58. G.M. Van Dyne, "Ecosystems, Systems Ecology and Systems Ecologists", in B.C. Patten and S.I. Jorgensen, eds, *Complex Ecology*, (NJ: Prentice-Hall, 1995), p.5.
59. Goldsmith, *The Way*, p.112.
60. See the following: C.J. Glacken, *Traces on the Rhodian Shore: Nature and Culture in Western Thought From Ancient Times to the End of the 18th Century* (Berkeley: University of California Press, 1967), F.N. Egerton, "Changing Concepts in the Balance of Nature", *Quarterly Review of Biology*, 48, (1973):322-350; D.B. Botkin and M.J. Sobel, "Stability in Time-Varying Ecosystems", *American Naturalist*, 109, (1975):625-646; D. Botkin, *Discordant Harmonies: A New Ecology for the 21st Century* (Oxford :Oxford University Press, 1990), D. Worster, *Nature's Economy: A History of Ecological Ideas* (San Francisco: Sierra Club, 1994).
61. Specific outbreaks of ecological imbalance at various scales include the following :

    A) At the local level, outbreaks of predation in which defoliation exceeds 100%, such as the catastrophic changes to fields and corals caused by local population outbreaks of plague caterpillar (see G. Conway, "Man Versus Pests", in R. May, ed, *Theoretical Ecology: Principles and Applications*, Oxford: Blackwell, 1976) and the crown of thorns starfish (see A.M. Cameron, and R. Endean, "Renewed Population Outbreaks of a Rare and Specialized Carnivore, the Starfish *Acanthaster plancii*, in a Complex High Diversity System (the Great Barrier Reef)", *Proceedings of the 4th International Coral Reef Symposium*, 1982:41-49).

    B) At the regional level, the lack of stability in North American forest over ecological time as well as over geological time (see G.F. Peterkin and C.R. Tubbs "Woodland Regeneration in the New Forest, Hampshire Since 1650", *Journal of Ecology*, 2, 1965:159-170; D.B. Botkin and M.J. Sobel, "Stability in Time-Varying Ecosystems",

1965:159-170; D.B. Botkin and M.J. Sobel, "Stability in Time-Varying Ecosystems", *American Naturalist*, 109, 1975:625-646), M.B. Davis, "Quaternary History and the Stability of Forest Communities", in D. West *et al*, eds, *Forest Succession: Concepts and Applications, Springer-Verlag*, Berlin, 1981); H.R. Delcourt *et al*, "Dynamic Plant Ecology: The Spectrum of Vegetational Change in Space and Time", *Quaternary Science Review* 1, 1982:153-176.)

C) At the global level, global mass extinctions in the Cambrian, the Ordivician, the Permian, the Cretaceous, and throughout the Eocene can be regarded as cases of ecological imbalance in ecological collections of global and evolutionary importance.

Further general studies (including theory proposals) of biotic imbalance in ecological collections can be found in: F.N. Egerton, "Changing Concepts in the Balance of Nature", *Quarterly Review of Biology*, 48, (1973):322-350; J.H. Connell and R.O. Slatyer, "Mechanisms of Succession in Communities and Their Role in Stability and Organisation", *American Naturalist*, 111, (1977):1119-1149; P.S. White, "Pattern, Process and Natural Disturbance in Vegetation" *Botanical Review*, 45, (1979):229-299; J.H. Connell and W.P. Sousa, "On the Evidence Needed to Judge Ecological Stability and Persistence"; J.A. Wiens, "On Understanding a Non-Equilibrium World: Myth and Reality in Community Patterns and Process", in D.R. Strong *et al*, eds, *Ecological Communities: Conceptual Issues and Evidence* (Princeton; Princeton University Press, 1984):439-457; S.T.A. Pickett and P.S. White, eds, *The Ecology of Natural Disturbance and Patch Dynamics*, (Orlando: Academic Press, 1985); P.L. Chesson and T.J. Case "Overview: Non-equilibrium Community Theories: Chance, Variability, History and Co-existence", in J. Diamond and T.J. Case, eds, *Community Ecology* (NY: Harper & Row, 1986): 229-239; R. May, "The Search for Patterns to the Balance of Nature: Advances and Retreats", *Ecology*, 67, (1986):1115-1126; J. Wu and O.L. Loucks, "From Balance of Nature to Hierarchical Patch Dynamics: A Paradigm Shift in Ecology", *Quarterly Review of Biology*, 70, (1995):439-466. For attempts to alert environmental readers of such ecological imbalance see: A. Brennan, *Thinking About Nature*, University of Georgia Press, Athens, (1988). H. Cahen, "Against the Moral Considerability of Ecosystem", *Environmental Ethics*, 10, (1988) :195-216; D. Botkin, *Discordant Harmonies: A New Ecology for the 21st Century* (Oxford: Oxford University Press, 1990); D. Worster, "The Ecology of Order and Chaos", in S.J. Armstrong and R.C. Botzler, eds, *Environmental Ethics: Divergence and Convergence* (NY: McGraw-Hill, 1993) pp.39-48. Borrowing a passage from the environmental historian, Donald Worster, it might be appropriate to consider that:

> "the message [in all these works]…is consistent: the old ideal of equilibrium is dead; the ecosystem has receded in usefulness; and in their place we have the idea of the lowly 'patch'. Nature should be regarded as a landscape of patches of all sizes, textures, colours, changing continually through time and space, responding to an unceasing barrage of perturbations." (Quote from D. Worster, "Nature and the Disorder of History", in M.E. Soule and G. Lease, eds, *Reinventing Nature: Responses to Postmodern Deconstruction*, Washington, DC: Island Press, 1995, p.74.)

62. For examples of publications by those who adhere to this view see: D. Thistle, "Natural Physical Disturbance and Communities of Marine Salt Bottoms", *Marine Ecological Program Series*, 6, (1981):223-228; H. Caswell, "Life History Theory and the Equilibrium Status of Populations", *American Naturalist*, 120, (1982):317-339; Connell and Sousa, "On the Evidence Needed to Judge Ecological Stability and Persistence", G.P. Harris, *Phytoplankton Ecology: Structure, Function and Fluctuation* (London: Chapman and Hall, 1986), *Burrows, Processes of Vegetation Change*.
63. Wiens, "On Understanding a Non-Equilibrium World: Myth and Reality in Community Patterns and Process"
64. Connell and Sousa, "On the Evidence Needed to Judge Ecological Stability and Persistence" p.794.
65. P.W. Frank "Life Histories and Community Stability" *Ecology*, 49 (1968):355-357.
66. F. Capra, *The Web of Life* (NY: HarperCollins, 1996) p.293.
67. H.O. Hallman, "Nietzsche's Environmental Ethics", *Environmental Ethics*, 13, (1991):99-125.
68. See J.B. Callicott, *In Defence of the Land Ethic: Essays in Environmental Philosophy* (Albany: SUNY Press, 1989).
69. Burrows, *Processes of Vegetation Change*, p.466.
70. G.C. Williams "Gaia, Nature Worship and Biocentric Fallacies" *Quarterly Review of Biology*, 67 (1992):470-483.
71. Wu and Loucks "From Balance of Nature to Hierarchical Patch Dynamics" p.442.

CHAPTER 3

Gaia:
The Technocentric Embodiment of the Unity of Nature?

Introduction: Gaia--The Ultimate Organism and the Ultimate Ecosystem

The story which is recounted in Chapter One indicates that holistic, organic theories of nature are inherently ecological, inherently able to offer valuable stories about how to save the planet from destruction. Gaia theory might be regarded as the Earthly apex of such holistic, unitarian views of nature.[1] However, the Gaia theory might not be as environmentally friendly as many of its supporters may think, and in this chapter it is indicated how Gaia may be a theory just as endowed with ecological fascism as the superorganismic and systems ecology theories addressed in the previous chapter. This chapter also erodes the environmental potential of the Gaia theory by suggesting it is a theory of high technology, a theory of imperialism and a theory that is hostile to socially-aware environmental policy-making.

The Gaia theory was initially conceived by James Lovelock in the 1960s and developed by him and a handful of other scientists, most notably Lynn Margulis, throughout the 1970s, 80s and 90s. The theory's title comes from the ancient Greek Goddess of the Earth who went by the same name.

The Gaia theory states that the whole of the world's biota constitutes one great big homeostatic system; a system that is palpably analogous to the system that is a cell, an organism or an ecosystem. Like cells, organisms and ecosystems the global system is united by the internal flow and cycling of matter and energy within the system.

In recent years, there has been a concerted effort to claim that the Gaia theory is the modern incarnation of the superorganismic thesis that Clements promoted earlier in the century. For instance, the second Oxford University gathering of Gaian specialists had as its main theme: 'the Evolution of the Superorganism'. Similarly, in an article in the *New Scientist*, Roger Lewin reaffirms the existence of Gaia under the specific intellectual framework of the 'superorganism'.[2] These acknowledged links between Gaia and superorganicism are of telling importance and will be examined in this and other sections.

So what is the Gaia theory, exactly? According to Bonsor:

> simply stated the Gaia hypothesis suggests that life as it exists today on the Earth does so because in the distant past early life created conditions suitable to itself, and that since that time life, in concert with some purely chemical and physical processes of the planet, has been actively maintaining those conditions at optimum values.[3]

Although universal scientific respectability for the theory is not entirely forthcoming there is nevertheless a lot of support for Gaia from those scientists with a penchant for superorganisms and systems thinking. The systems ecologists Eugene Odum and Pat Klinger, for instance, are on record as supporting the Gaia theory[4]. Perhaps more importantly, however, is the support given to the Gaia theory by numerous environmentalists. Edward Goldsmith for instance praises the global systems science of James Lovelock:

> whose Gaia thesis I (and a lot of other people) regard as absolutely vital to the development of an ecological worldview.[5]

Similarly, David Abram, Michael Allaby, Theodore Roszak, and Carolyn Merchant are amongst the many environmentalists who lavish praise upon the theory for it finally provides scientific proof of what they had known all along: nature is in unity.[6]

For Edward Goldsmith, Gaia has become the fundamental unit of study in ecology ("Ecology studies natural systems in their Gaian context", says Goldsmith) as well as the fundamental unit of study in evolutionary biology ("Gaia seen as a total spatio-temporal process, is the unit of evolution" declares Goldsmith).[7] Gaia is not only the fundamental unit in biological study, however. It is, according to Goldsmith, fundamental in social studies, too, as we shall see later.

One of the selling points of Gaia for Lovelock, and many others that write about the theory, is its presumed non-anthropocentric framework.[8] Gaia it is claimed, transcends anthropocentrism by detailing exactly how the systems of the Gaian organism are not attuned to caring about human aspirations and endeavours. If humanity hurts and hinders Gaian processes it will not necessarily be Gaia that suffers but humanity since Gaia may re-adjust its systems to exclude humans by altering the precise ambient conditions of its physical environment.

Because Gaia is philosophically (and historically) linked with systems science and superorganicism it suffers from much of the same scientific problems that these intellectual traditions suffer from. For instance; just as it is difficult to confirm the stability of an ecosystem merely because it is in energetic and material equilibrium so it is difficult to confirm the stability of the whole planet on the same basis. Likewise the negative feedback patterns observable (to some extent) in local ecosystems might be observable in a global Gaian system but in both cases such feedback may only be associated with some minor characteristic of interest that has little or no relevance to the biotic components. Also, as environmental philosopher Andrew Brennan points out, such global feedback patterns may exist but this does not mean they are "of necessity complex and densely interrelated."[9]

As well as suffering from familiar scientific problems, Gaia theory also suffers from the same philosophical and ethical problems that systems ecology suffers from. Gaia functionalises, de-personalises and de-ecologises the natural entities of the world, converting them into typological black boxes that serve as mere conduits of energy and matter.

The intellectual linkage between Gaianism and the systems thinking of ecosystem ecology and systems ecology can be quickly summed up by examining the following sentence written by Lovelock:

> Because Gaia was seen from outside as a physiological system, I have called the science of Gaia geophysiology.[10]

Lovelock's focus on global, whole Earth physiology has led him to label his own peculiar brand of science 'geophysiology'. By assigning his theory to this new science (and indeed, by heralding Gaia as the flag-bearer of this new science), Lovelock then proceeds to offer himself up as a 'planetary physician' involved in the diagnosis and cure of the Earth's maladies.[11]

Lovelock's appeal to physiology is an echo of previous holistic scientists, of course. Just as Clements saw prairie communities as analogous to organismal physiological systems and just as the systems ecologists utilised physiological analogies to describe their ecosystems, so Lovelock, and his disciples, rely on physiology to define the particular subject of his focus; the Earth. The exact analogies are sometimes slippery. Gaia at times is compared to the physiology of a cell, then to the physiology of an organism or the human body[12], then to the physiology of a whole ecosystem. At all points, though, the primacy of looking at the

Earth as a physiological system generally remains. Really, it is relatively unimportant as to what metaphor is used to explain the functioning of Gaia since, to Gaians, all these biological entities function in the same way: as systems.

Gaia and Ecological Functionalism

As with the functionalism of systems ecology so, too, the functionalism of Gaian ecology declares that the whole is of more importance than the parts. In the case of Gaia, the whole equals the planet:

> Gaia theory forces a planetary perspective. It is the health of the planet that matters, not that of some individual species of organisms[13]

The real importance of the parts is derived from their role in contributing to the whole; and not all parts are of equal value in Lovelock's Gaia. Some parts of the geophysiology of Gaia are declared by Lovelock as essential and some are considered expendable:

> Gaia has vital organs at the core, as well as expendable or redundant ones in the periphery. What we may do to the planet may depend greatly on where we do it.[14]

Through this differentiation of importance, some species and ecosystems are devalued because they do not significantly contribute to the matter and flows that make up the Gaianl system. For instance, while coastal wetlands and tropical forests are identified by Lovelock as being essential to the cycling of matter and energy in Gaia's geophysiology, many other types of ecosystems (for instance Arctic tundra and benthic ecosystems) are superfluous redundancies which may be forfeited without consequence to the Gaian superorganism. This sort of variable valuation of the Earth's ecosystems pervades Lovelock's work. For instance he also says:

> dense perturbation in the northern temperate regions may be less a perturbation than in the humid tropics.[15]

The emphasis upon functions and roles within Gaia theory leads Lovelock to rather startling conclusions with regard to the destruction of species and the despoliation of large stretches of the Earth's environment. Gaian ecology suggests, for example, that in the event of massive changes in the taxonomic composition of the Earth's biota, the identity of Gaia remains unchanged because the mechanisms involved in the geophysiological processes of Gaia (i.e. matter and energy cycling) remain in place. The extinctions of the Cambrian Period and during the last ice ages could not have impacted upon Gaia's identity, it is said by Lovelock, since: "An ice age doesn't really seriously affect the whole Earth as a living system" because the matter and energy cycles recover quickly in the long term scheme of things to become stable again.[16]

This sort of long-termism and predilection towards grand scales is also present in Lewin's recent defence of the importance of the superorganism concept. He states:

> In the history of the Earth since complex forms of life evolved 500 million years ago the biota has collapsed five times in mass extinctions involving the loss of at least 50 per cent of species and sometimes as much as 95 percent. After each extinction the biota bounced back in diversity with different players on the stage, but with the pattern of interaction--the pattern of niches--remaining the same. Reptiles were the major carnivores before the extinction at the end of the Cretaceous 65 million years ago, and mammals filled that role afterward for instance.[17]

The extinctions Lewin talks about could possibly be interpreted as manifest examples of difference (since wholly different types of organisms are given rise to) but they are not. They are interpreted--via 'roles' and 'functions'--as constituting sameness and changelessness.

This emphasis on roles and functions rather than those entities that undertake them is potentially troublesome. If all things are converted to roles, the concept of role performance becomes more important than the actual individuals and species who supposedly perform them. Apart from being ethically dubious, since individuals become entirely associated with their roles and nothing else, this is also ecologically dubious. Individual organisms drift away from and reject their supposed ecological roles as it best suits them. Andrew Brennan alludes to this phenomenon in his statement about how:

> many organisms are ecological opportunists, exploiting what resources they can and switching their feeding preferences according to the resources available and the competition they face.[18]

The expression of the idea of essential ecological roles for species/organisms is most manifest in the ecological concept of the 'niche'; a term that is often interpreted to mean the ecological profession or job of an individual species/organism. The concept of the niche, however, may be a tyrannous one since when using it, the ecological reality and environmental identity of a species/organism is directly correlated to its prescribed job. If environmental values are drawn from this, it will lead to a presumption of superior value in niche fulfillment rather than the intrinsic value of the species/organism itself.[19]

Lovelock both tends towards and extends the tyranny of the niche by applying niche roles not just to organisms and species but to entire ecosystems and biomes. Thus a tropical rainforest is worthy of preservation, according to Lovelock, not because it is intrinsically valuable or because it houses intrinsically valuable species and individuals but because it is functionally important to Gaia:

> we are failing to recognise the true value of the forest, as a self-regulating system that keeps the climate of the region comfortable for life.[20]

For many environmentalists, the emphasis on functions and roles within Gaia theory, which we may refer to as Gaian functionalism, is a source of strength for the politics and philosophy of environmentalism since they can use it to say that people and governments cannot go about hurting ecosystems haphazardly since the whole biosphere might collapse into an undesirable state.

However, perhaps such sentiments are not best expressed by using Gaia's brand of functionalism but via alternative brands of ecological functionalism. For instance, it might be useful for environmentalists to compare Lovelock's Gaian functionalism (with its differential ascription of what is ecologically useful/useless) to the ecological functionalism of someone like Rachel Carson, who in the early 1960s promoted a different brand.[21] Firstly, Carson was concentrating on the ecosystem level of nature and nothing so grand as the whole planet; and secondly, her work suggested that all organisms and communities were essential for the normal functioning of the environment. Unlike Lovelock's Gaianism, this meant humanity had to protect all types of ecosystems and all species rather than those deemed by humans to be of essential function. What's more, in Rachel Carson's functionalist ideas, we did not know what the consequences of decimating ecosystems and species would be. Science can not judge, for example, what may happen if tropical rainforests were cut down.

Compared to Carson's ideas, Lovelock is much more omniscient. Not only does he know the processes that go on but he can actively identify those that are redundant. Whereas Carson would have said 'don't chop down the tropical rainforests and don't destroy the benthic ecosystems of the Earth because we don't know what the ramifications for the rest of the world will be', Lovelock would say 'don't chop down the tropical rainforests because that will upset the atmospheric chemistry of the world but trawl and despoil benthic ecosystems all you like because they don't do anything important'. No new scientific evidence has been found since Carson's time for Lovelock to arrive at his conclusions; the difference is merely one of philosophical perspective. Lovelock is prepared to declare that weird things happen but science can sort them out, whereas Carson declares weird things happen but science cannot follow them and cannot sort them out.

When it comes to addressing those ecological scientists who believe that the world is not nearly as stable and continuous over space and time as his Gaia theory suggests, Lovelock brings in some telling analogies. Theoretical ecology, Lovelock says:

> is more concerned with sick than healthy ecosystems. The vagaries of weather are more interesting than the long-term stability of climate.[22]

Here the Gaian bias of regarding change as a symptom of sickness, while balance and stability are regarded as healthy, is clearly expressed. Ecosystems that change are unwell according to Lovelock; ill, abnormal, and not worthy of generalising from. In this way Lovelock castigates those things which are inherently changeable as of peripheral concern. What this means from an environmental and ethical point of view is that what we should be protecting and striving for are ecological settings that remain the same.

From Lovelock's Gaian point of view, it seems there is a certain way that things behave; a normal 'stable' way, and any deviation from this is abnormal. This idea also emerges strongly from some environmental philosophy. For instance, going back to Westra's concept of integrity: if "integrity conveys the idea of wholeness and of unbroken functioning," then, according to Westra's philosophy, disintegration conveys the idea of dysfunction and separation.[23] Here we come to see a particular intellectual fixation within unitarianism emerging. If things are not unified, they are broken, disintegrated, and dysfunctional.

Gaia and Physical Reductionism

Lovelock and Margulis' scientific conceptualization of Gaia theory runs like this:

> The Gaia hypothesis states that the surface temperature, composition of the reactive gases, oxidation state, alkalinity-acidity on today's Earth are kept homeorrhetically at values set by the sum of the activities of the current biota.[24]

It might be asked how come Lovelock and Margulis find it so easy to jump from the self-regulation of the world's chemistry to thinking of the world as a living organism? It seems somewhat of a long bow to draw to claim that the biotic control of the Earth's ocean and atmospheric chemistry must lead to the conclusion that the Earth is a living organism. Not for Lovelock (and other Gaians), however. For them it is easy, since, from the start, their view of the Earth's biota, and of living things, is a physico-chemical one. Life is directly equated with physico-chemical homeostasis so that anything displaying physico-chemical homeostasis is alive.

That Lovelock (and other Gaians) are conducting a physico-chemical interpretation of life is exemplified in Tyler Volk's book about Gaia when he declares that the prime directives of the Gaia theory are: 1) "Attend to the cycles of matter", and 2) "Attend to the cycles of causes (of the cycles of matter)."[25]

Under Gaia theory, individual species and ecological communities are, first of all, physical entities whose existence is defined by their chemical consumptions and productions. This implies that a plant, to Lovelock and Volk, is mainly a photosynthesis factory; a collection of mitochondria and chloroplasts that fix carbon dioxide and release oxygen. Like systems ecology, Gaia theory claims to be holistic. But by reducing life to physical essences it is actually reductionist.

If the Earth emerges, through Lovelock and Volk worldview, as a global unity whose parts are united by the passing of matter and energy between them, then it seems as though a person is at one with nature not because he or she may care for it but because they transfer matter and energy with it; i.e.: because they eat it and it eats them.

As with many systems views of the ecological world, the production and consumption of food is often taken as the basic interaction of nature. In ecosystem parlance, and as most high school science students are aware, all living things are categorised according to their role in the production or consumption of food; plants being labeled 'producers'; animals being labeled 'consumers'; while fungi and bacteria are given the distinction of being the world's 'decomposers'. This fetishism for food production and consumption contributes to the tyranny of the niche spoken about above since species are categorised as being in a particular niche according to what they do with their food. Brennan, in disagreement with such a view, comments that:

> if things were entirely what they ate, then the story about the natures of things would be less complicated than it is.[26]

This objection by Brennan is not the only philosophical objection that can be made towards the production and consumption fetish of the Gaian systems approach. At the global scale, according to Gaianism, it seems as though the Earth's various production and consumption components may rightly be visualised as unified masses. For instance, whereas systems ecology converts all plants into singular producers, Gaia theory would go further and present all the world's plants as part of the "Global Green Photon Harvester."[27] Such choices of metaphor further expose Gaia's overlording holism since the unique behaviour and values of individual plants are glossed over as they merge together to become faceless functional components in the great Gaian entity.

Gaia and Environmental Policy

One might believe that the functionalism and overlording holism within Gaia is hardly about to affect any real concrete environmental situations that the Gaia theory gets itself involved with. This may not be the case, however. For one thing, it is clear that Lovelock thinks that the Gaian organism can just about take care of herself and that many environmental programs are in error because they do not take this into account.

Gaia's ability to regulate herself via intricate feedback mechanisms means, says Lovelock, that the Earth's biota is never fully in danger. True, he admits, human activities might produce feedback within the global system that eventually produces a world unpalatable to human concerns but life on Earth as a whole will never be destroyed.

At pains to expose what he thinks is an environmental myth, Lovelock presses forward the view that he in no way considers life to be vulnerable or fragile. To enable himself to do this, Lovelock first converts 'life' (which we may signify with a small 'l') into the concept of 'Life' (which we may signify with a with a big 'L'). Whereas 'life' is open to all sorts of diverse scientific meanings and biological interpretations—from the selfish struggles of DNA molecules through the daily socio-psychological experiences of individuals to the fecund beauty of vegetated landscapes, Lovelock's 'Life' is always a total systems view of life; life at the global level. In this regard Lovelock believes that Life as a whole is hardly vulnerable--even though the lives of individuals and landscapes obviously are.

If Life is not vulnerable or fragile, then it makes the protestations of environmentalists seem a little strange. Life cannot be destroyed or harmed, so what is the use of worrying about those things that seem to destroy or harm it? Lovelock has always announced that he considers environmentalists to be a little skewed when voicing their worries since it is hardly Life that suffers from environmental problems, only humans. In fact:

> James Lovelock has suggested that the Green movement with its opposition to all industry and its contempt for science and technology, has been a potent force preventing the requisite environmental reform.[28]

This last statement shows up a key political strand of thought within Lovelock's environmental credentials. Firstly; he adheres to a brand of environmentalism that most environmentalists would call 'technocentrism'. David Pepper defines technocentrism as a type of environmental thought:

> which recognises environmental problems but believes either unrestrainedly that man will always solve them and achieve unlimited growth...or, more cautiously, that by careful economic and environmental management they can be negotiated...In either case considerable faith is placed in the ability and usefulness of classical science, technology, conventional economic reasoning (e.g. cost-benefit analysis), and their practitioners.[29]

Thus, says Lovelock, scientific know-how and technological growth can be an intimate part of understanding and protecting Gaia. In fact in *The Ages of Gaia* and *Healing Gaia* he actively encourages industrial growth since it provides the intellectual and economic impetus to solve environmental problems. This would stand in sharp contrast to many environmentalists who see both science and technology and unrestrained economic growth as major contributory factors to the environmental crisis.[30]

Lovelock is also negatively disposed to the social relevance of environmental ills. This is evident, for instance, when Lawrence Joseph recounts an anecdote that he gained from Lovelock himself. Conversing at a dinner table at an international development-environment meeting we find both Lovelock and Mother Theresa. According to Joseph, this is the essence of the conversation:

> Their differences were clear. Essentially, Lovelock's message was: 'take care of the Earth and humanity's problems will start to take care of themselves'. Mother Theresa expressed just the opposite: 'take care of the people, and the Earth's problems will come round'.[31]

Lovelock's asocial approach to environmental problems is characterised most saliently with regard to population issues. When discussing population issues he chooses to describe the human population increase as a 'serious planetary malady'. If there were one environmental policy that Lovelock would like to implement upon the world (if he had the power) it would be population control. Befitting someone who confides in nature's way of doing things, Lovelock gains his knowledge of population issues not from the social sciences but from the natural and mathematical sciences as he goes ahead to compare human population growth with lilies in a lily pond:

> To understand what exponential growth is really like, imagine a lily pond on whose surface a water lily is growing and spreading so as to double its leaf area daily. It has taken the lily 19 days to cover half of the pond's surface with its leaves. How long will it be before the leaf area doubles again so that the pond is entirely

> covered? Does the answer come instantly to mind? Not another 19 days, but one day.[32]

Lovelock then goes on to assert that: "We humans must be close to that 19 day period of lily growth"[33]

It might be contended, however, that unlike lily ponds, humans tend to reproduce more when resources are low since there is a lack of reproductive freedom and a lack of contraception available to those who tend to reproduce the most (poor rural women of the developing world) and also because high reproductive rates, i.e.: having many children, is often the only way of insuring that the parents are provided with welfare in old age. Social relations that lead to these circumstances are hardly related to the mathematics and biology of lilies and their ponds. They are more to do with resource maldistribution amongst rich and poor sections of society, and amongst First and Third World nations.[34]

Lovelock's naturalisation of the problem of overpopulation avoids the social background that leads to it. This, in effect, would seem to encourage the population problem, about which Lovelock warns his readers, since it deflects any understanding of how we should solve the problem away from the social circumstances that must be sorted out in order for the problem to be addressed properly.

Where Lovelock envisions the population problem to be at its worse is also very telling. For Lovelock, the worst concentrations of the population problem are in the tropics; in those nations that are home to the tropical rainforests which Lovelock holds to be the 'air-conditioning units of Gaia'. Temperate areas, being not so essential for the Gaian organism, are not so impacted by high population numbers: "dense population in the northern temperate regions may be less a perturbation than in the humid tropics".[35] If one wants to be cynical about this perspective, one could point out that Lovelock seems to want to kill off the poor dark people of the world in order to save it for a few wealthy white people. Lovelock does not say this, of course, but the intimations are there in *Healing Gaia* and *The Ages of Gaia* that population control can be distributed with differing intensities in different regions. This varying distribution of population control might easily be convertible into unjust population plans by future Gaian practitioners.

The depth of Lovelock's anti-population feelings are revealed when he says: "The statement 'There is no pollution but people' carries an awful truth."[36] He goes on to say:

> None of the environmental agonies now confronting us--the destruction of the tropical forests; the degradation of land and seas; the looming threat of global warming; ozone depletion and acid rain--would be a perceptible problem at a global population of 50 millions.[37]

A less technocentric environmentalist might indicate that at any population, humanity's current fixation with big technology and capitalist industrialism would reek havoc upon the world's environment. If humanity somehow did manage to decrease the population of itself down to the fifty millions, consumerism, technology and industrialism would just expand so as to effect the damage that was previously wrought by six billion people. In any case, Lovelock's identification of overpopulation as the key environmental malady along with his encouraging appraisal of industrialism, the 'technological fix', and economic growth would surely make most environmentalists think that he was quite incapable of appreciating and understanding the various social and political causes of the environmental crisis.

Another aspect of environmental policy that emerges from Gaia, and in particular from Lovelock's Gaia, is its soft approach to pollution control. For Lovelock, the 'poison is the dose'. What he means to indicate by referring to this notion is that industrial pollution is only bad when there is too much of it. This seems to be a reasonable enough assertion but if we take his definition of 'too much' we find that it revolves around the protection of Gaia's energy and matter cycles. These, Lovelock says over and over again, are very easy to protect (in fact impossible to destroy) even if industrialism, technology, consumerism and pollution is rampant in the world. Thus Lovelock finds it easy to support uranium mining, nuclear power, and various forms of techno-industrial growth because they hardly, according to him, affect Gaia's ability to operate her self-regulating feedback mechanisms.

If we base the definition of 'too much' on less abstract notions than Life, Gaia or the Global System we find that Lovelock's acceptable doses are much higher than those acceptable to local ecological communities. It is of no consequence for Gaia (and Lovelock) if a nuclear meltdown occurs in the Ukraine or if an oil spill occurs on the Alaskan coast since, although the local environment is destroyed, the environment as a global whole carries on in the form of the abstract Gaia. Similarly, the lives that are destroyed in these disasters carry on in the form of Lovelock's abstract notion of 'Life'. (Lovelock would also state, of course, that the oil and nuclear fuel industries that initiated these local disasters actually operate in the long run to save the environment by providing economic growth and thus stimulus for technological innovation and scientific know-how which could be used to fight further environmental problems).

All this said, many might still argue that the Gaia theory has at least enabled us to see the interconnections between concrete living systems. However, there was a long enough tradition in ecology for this to be the case without Gaia being around (as has been explored in Chapter Two. Another related argument would point out that the Gaia theory, as a philosophy, makes us aware and conscious of vital processes that must be protected. But as Andrew Brennan points out:

> We do not need to imagine Gaia to be a unified superorganism to appreciate the importance of regulating the salinity of the sea, the oxygen content of the atmosphere, the volume of the ozone layer and so on.[38]

The Selling of Gaia as Science and Philosophy

The notion that the Earth has planetary-wide physical and geochemical cycles is actually not James Lovelock's invention. Alfred Lotka in the early decades of the Twentieth century, for instance:

> viewed the Earth as a single system linked by exchanges of chemical elements and driven by solar energy.[39]

Hagen also finds that:

> The idea that the chemistry of aquatic ecosystems is regulated by organic activities had also been briefly discussed by the biological oceanographer Alfner Redfield during the late 1950s.[40]

Lovelock, himself, also delights in drawing attention to James Hutton's superficial ruminations of the Earth as a living being. (Lovelock does this to present himself as amongst distinguished scientific company.[41])

If there are all these progenitors to Gaia why did they not attach philosophical importance to it like Lovelock? In short, they probably did. Hutton adapted his geo-organism to conform with the geological uniformitarianism he adhered to, and Lotka fitted in his physically-unified Earth to the quantitative predilections of his quantitative biology.[42] Lovelock, likewise, was operating within the worldview of mid to late Twentieth century systems theory but he was able to jump on the political acceptability associated with the currently fashionable penchant towards holistic ecology in order to sell his theory both within science and outside of it.

We may still ask, however, why Gaia has become so popular as an ecophilosophy in public circles and not, say, Clements' superorganism. Apart from Lovelock's obvious predilection toward popularising his ideas we might indicate here a preoccupation in the late Twentieth century public for 'worldism'. The icon that many environmentalists often refer to and revere is not the forest, the lake or the mountain but the Earth as a whole. This reverence is best encapsulated within the iconographic photos of the planet Earth as taken from space. This abstract image of the Earth floating as an island in space acts as an all-encompassing symbol of the environmental movement. It not only symbolises the fragility and loneliness of the

Earth but it also symbolises everything which is at stake since the Earth contains everything we know that is alive. The adoration of the Earth image thus reflects a holistic mindset which must be attended to in order for the Earth to be preserved. The Earth image becomes the ultimate embodiment of holistic thought since it is the greatest holistic perspective that can be taken when dealing with environmental issues. This is where Gaia may gain a lot of its selling power.

That such mixtures of philosophy and science pervade the cultural acceptance of Gaia should not be too much of a surprise since Lovelock, Margulis and other scientific Gaians are themselves involved in selling Gaia not just through science but through philosophical premises as well. This can be observed, for example, in the following passages:

> We inhabit a global metabolism with a four-billion year pedigree. In just the past few decades, we have awakened to an awareness of damage we ourselves are inflicting on this metabolism with our blind urges to procreate and appropriate. Perhaps in the fantasies of many, the 'taxi outta here' would be a time machine. Just tell the time-taxi to stop at nature a thousand or more years ago. But because most of us would want to keep the postal service, the Internet, MRI scans, and an abundance of items in stock at the local supermarket, we must proceed with the world as it is, and that requires knowing how the foundational processes of nature work. I am convinced that such knowledge, if widely held, will contribute to shaping the future of Gaia--a future in which we, as a new biochemical guild, will necessarily be integrated into the global metabolism, for better or for worse.[43]

> The essence of living green, of being a citizen of Gaia is not fretful puritanism. If we can think of ourselves as part of a giant living organism and perhaps even a cause of its indigestion, then we may be guided to live within Gaia in a way that is seemly and healthy. Even thinking this way is an antidote to the fatalism of accepting the Earth as dead with life as just a passenger.[44]

While employing such philosophical appraisals of nature and humanity to sell the Gaia theory, Lovelock, Margulis, Volk and the other scientific Gaians go to great lengths to spell out that their theory is a theory embedded in science. For example Lovelock declares that Gaia is not just a metaphor but is hard-core science: "I insist that Gaia theory itself is proper science."[45]

If Gaia is a philosophy of nature or a metaphysical viewpoint, then, according to Lovelock, Margulis and Volk, it is one intensely devoted to scientific ideals. Those people that adopt the philosophy and metaphysics of Gaia without its science have been known to offend the scientific Gaians. This is most evident when Lovelock and Margulis' Gaia is usurped by the philosophy of certain environmental and New Age spiritual movements:

> This is where Gaia and the environmental movement...part company.[46]

> The religious overtones of Gaia make me sick.[47]

Such responses are directed by Lovelock and Margulis toward a myriad of ecophilosophers, spiritual ecologists and Gaia popularisers who tend to utilise Gaia theory to prove the sanctity or divinity of the world and the value of all things in it.

While they, themselves, enthusiastically philosophise about the meaning of Gaia, both Lovelock and Margulis have shown a certain intolerance towards certain Gaian philosophising that they do not agree with. The interesting thing is that the first thing they tend to do when presented with distasteful philosophies of Gaia is to come over all hyper-scientific, retreating into the authority of science in order to bolster their defences from within its walls.

What is also interesting is that when the Gaia theory is criticized from a scientific viewpoint, Lovelock and Margulis often resort to philosophical premises to gain ground. For instance, Lovelock regularly revises Gaia in different philosophical terms so as to get the Gaian message across in a slightly different way (for example, Lovelock defends his idea that the Earth is alive by variously relating it to living trees, self-regulating machines and cybernetic processes, depending on who he is responding to).

This does not mean that Lovelock has not tried to enter into a 'scientific' debate with his scientific critics but he only does this if he considers the critique worthy enough to rebuke. Or, perhaps more accurately, he only does this when the criticism can be rebuked within the scientific background presupposed by Gaia theory (for instance, Lovelock's creation of the Daisy World model[48] arose out of his acceptance of the anti-teleological arguments of some scientific Gaia critics).

For those scientific criticisms of Gaia that can not be rebuked through Gaian science, Lovelock also has a strategy. He may claim that such criticism is, itself, unscientific. For instance, when addressing the criticism that Gaia theory is unscientific Lovelock claims that those who level such criticisms are themselves operating unscientifically. This is evident in the following paragraph:

> creative science is the province of working scientists, and few of these are found among our critics. Those who condemn Gaia as unscientific are, for the most part, science writers and professional science critics. Creative science suggests experiments. This is something that science writers and critics rarely wish to know, for it would make the telling of the stories so much more difficult.[49]

Here Lovelock makes the declaration that if you claim Gaia as unscientific you, yourself, are not scientific and so, conversely, he is making the claim that the only scientific approach to take is to believe in Gaia.

What seems to be happening in all this discursive articulation between the philosophy of nature, ecophilosophy and science is the constant self-positioning of Gaia theory within the higher bastions of human knowledge. Both Lovelock and Margulis are aware of the great prestige and status of science and both are aware that they have to operate within it when making generalisations of the nature of the planet. They are also aware, though, that to sell it to both scientists and non-scientists you often have to describe the science in philosophical terms.[50]

Gaia and Non-Anthropocentrism

James Lovelock makes much of the non-anthropocentric focus of Gaianism. He states that Gaia is not a human-centred environmental view because it cares more about the health of the Earth than it does about the health of people and in doing this it differs radically from most conventional environmentalisms. For instance, in *Healing Gaia* he states:

> Environmentalists, churches, politicians, and science, all are concerned about the damage to the environment. But their concern is for the good of humankind. So deep is this introspection that even now few apart from eccentrics really care about other living organisms. The oft-stated objection to the rape of the forests is that they might include within them some rare plant that bears the cure for cancer, or that the trees fix carbon dioxide, so that if they are cut down we may no longer enjoy our privilege of private transport.

He repeats himself later in the same work when he states:

> Scientists and campaigners in the First World and in the tropics still try to plead for the preservation of these forests on the feeble and human-centred grounds that they are home to rare species of animals and plants, and especially to plants containing drugs that could cure cancer and other frightening human diseases.[51]

This is a line Lovelock peddles in all of his three major Gaia books; from *Gaia: A New Look of Life on Earth*, through *The Ages of Gaia*, to *Healing Gaia*. One notable thing about these statements is the profound ignorance Lovelock seems to exhibit with regards to the non-anthropocentric nature of many (probably most) contemporary streams of environmentalism. Deep Ecology, Third World ecology, ecofeminism, ecosocialism, bioregionalism and Social Ecology (let alone those environmentalists involved in scholarly fields such as environmental philosophy, environmental sociology, literary ecocriticism and environmental economics) have all entered into a debate about anthropocentrism which is considerably more sophisticated than the 'preserve nature for humans' line that Lovelock thinks environmentalists uniformly adopt.[52] In just about any book of recent environmental scholarship dealing with values, this 'preserve nature for humans' anthropocentrism has been thoroughly questioned, if not rejected, and the protection of non-human species and non-human environments is now commonly advocated upon the intrinsic value of these species and environments. What can be demonstrated also is that Lovelock himself--despite warning against human-centredness and accusing environmentalists of it--surely does not adhere to the anti-anthropocentric ideas that he suggests should be appreciated by all environmental sympathizers. This can be starkly demonstrated by quoting the sentences made by Lovelock in *Healing Gaia* that appear immediately after the passages quoted above:

> ...other frightening diseases. They may be so but they offer so much more than this. Through their capacity to evaporate vast volumes of water vapour the forests serve to keep their regions cool and moist, by wearing a white sunshade of reflecting clouds and bringing rain that sustains them. More even than this, the great forests of the tropics are part of the cooling and air-conditioning system of the whole Earth.[53]

> ..our privilege of private transport. None of this is bad only stupid. We are failing to recognise the true value of the forest, as a self-regulating system that keeps the climate of the region suitable for life.[54]

Here we see Lovelock is just as unable to recognise the intrinsic value of the animals and plants of the forests as the human-centred environmentalists he refers to, and he can only value them from a perspective of functionalism.

As well as accusing environmentalists of an anthropocentrism that few of them adhere to, Lovelock extols the Gaia theory as the very epitome of non-anthropocentrism for it is not the concern of Gaia to worry about humans. However, the crucial point here is that Lovelock does not need to extol the value of protecting the Earth for humans. He does not need to utilise anthropocentric arguments to look after the environment since his theory has been crafted and moulded into existence to do this implicitly. You do not need to promote an anthropocentric value system if your Goddess, by incredible good fortune, so willingly operates her laws to look after your interests. Gaia theory thus possesses a hidden type of anthropocentrism, an anthropocentrism that operates under the guise of Gaia but actually is just an incarnation of Lovelock's interests and philosophies. To me, it seems just too amazingly coincidental that a fan of scientific knowledge, worldwide industrialism, technological betterment and economic progress discovered a global Goddess that just happens to like these things too. Much more likely, is that Lovelock fashioned this Goddess to serve this purpose.

The Reproduction of Gaia: Lovelock's Technocentric Imperialist Organism.

According to Lynn Margulis, Dorian Sagan and Gregory Hinkle, the Gaia concept emerged from James Lovelock's fertile imagination and from the United States space program.[55] There is certainly a strong connection between space exploration and Gaia. Some of the most prominent Gaians are interested in space flight. For instance James Lovelock has been involved in the planetary program of NASA and has written a book about colonising Mars[56], Lynn Margulis was involved in NASA's Planetary Biology program and has also written about colonising Mars[57], Dorian Sagan, another keen Gaian, has also been known to write about space expansion[58], while Lewis Thomas, Freeman Dyson and Tyler Volk have written popular articles or books about both Gaia and space flight. Thomas has also been involved in NASA's Planetary Biology program, while Dyson is well-known for his grandiose space plans, and Volk has been involved in research devoted to working out the ecosystem properties needed for space colonies. It might also be noted that Rusty Schwieckert, an Apollo astronaut has also found Gaia theory of significant importance in his philosophical writings about space flight.[59]

For Lovelock specifically, the connection between space flight and Gaia theory is very strong since he acknowledges it as a prime factor in the theory's origin:

> For me, the personal revelation of Gaia came quite suddenly--like a flash of enlightenment. I was in a small room on top of a building at [NASA's] Jet Propulsion Laboratory in Pasadena, California. It was the Autumn of 1965, the room was an office of the Biosciences Division, and I was talking with a colleague, Dian Hitchcock, about a paper we were preparing on a method for remote detection of life on a planet.[60]

The planet about which Lovelock speaks is Mars and he goes on to explain that when he was working for NASA to investigate the possibility of life on Mars it suddenly dawned on him and his colleague that life on a planet could be detected remotely just by analysing the atmospheric composition from afar since an atmosphere created by a living planet would differ from the atmosphere of a dead planet. Thus if Mars was living it would have an atmosphere somewhat similar to the Earth's:

> Now the air of a dead planet would be expected to have a composition close to what is called the chemical equilibrium state. That is to say, all possible chemical reactions among the gases would have taken place and the atmosphere would be like the exhaust gas escaping from a car or a factory chimney; a mixture from which no more energy could be extracted. A planet that bore life would have a very different atmosphere because living organisms are obliged to use the air as a source of raw materials and as a depository for their waste products. Both of these uses would change the atmosphere away from the chemical equilibrium state. If the observed degree of disequilibrium among the gases of a planetary atmosphere was great, it would suggest the presence of life.[61]

In this regard Lovelock believed that the Martian atmosphere was in a state of chemical equilibrium[62], and so Mars is, according to Lovelock, quite dead.

This connection between space activities and Gaia takes an interesting turn when we note that James Lovelock, and the other above-named Gaians, are actually seeking to expand Earth life onto 'dead' planets such as Mars. Such expansion, to them, will constitute the reproduction of Gaia. The onward advance of humans into space is seen by these Gaians as indicative of the reproductive tendencies of the organism that is Gaia. Being an organism, Gaia is biologically inclined to reproduce. For instance, when summarising the Gaia theory for a space industry audience, Margulis says:

> The Gaia hypothesis states that the surface temperature, composition of the reactive gases, oxidation state, alkalinity-acidity on today's Earth are kept homeorrhetically at values set by the sum of the activities of the current biota. Life, in other words, is a planetary phenomenon. Since the natural tendency of all life is to grow exponentially to fill a proximal volume, a question for the 1990s is 'Can life expand to Mars?', that is, 'Can Gaia reproduce?'[63]

Martin-Smith, an advocate for human space expansion into the Solar System, can similarly justify his aspirations through Gaia theory:

> The Gaia theory has shown us how life and Earth have developed in lockstep, with life adapting Earth to suit its needs, via chemical homeostasis. This has provided a fruitful inspiration for the new devotees of the ancient Earth goddess, the environmentalists. Such a view, although poetic, is incomplete, since the prime role of Mothers is procreation! Slowly, the closed 'Spaceship Earth' ecological worldview is being challenged by a new paradigm--the fact that human ecology is Solar System-wide, and that our survival depends on extending our reach to match.[64]

Rusty Schweickart, ex-NASA astronaut, makes a similar argument for space expansion with regard to the Gaian organism:

> Humanity's accelerating growth rate and demands for resources and waste processing are the same kind of burgeoning needs that tax a mother's ability to nurture a growing foetus.[65]

No doubt Schweickart longs for the day when he can see Mars as a gently spinning green (instead of red) orb hanging serenely in the window of his spaceship, with blue mother Earth resting in the background for symbolic effect.

It is interesting to note that humans have a special role in this reproduction. They are either the midwives to the birth (as Martin-Smith, Sagan and Margulis, would state) or they are perceived as the seeds or spores of Gaia rightly acting to spread a representative biota of Earth to abiotic worlds (as Allaby would state[66]). These analogies, 'Gaia giving birth', 'humans as seeds', are analogies that seek to evade the social and political nature of human expansion into space. Those entities taking part in space expansion (like national governments, space agencies and aerospace companies) are social entities operating according to social, political and historical forces. Gaia theory, in this light, can be viewed is an attempt to naturalise the social, political and economic forces that lead to space expansion.

Michael Allaby indicates in his *Guide to Gaia* that humans might not literally be the seeds of the Earth but since we are behaving as though we were, it comes down to the same thing in the end. In this admission we see the strength of biological metaphors for Gaianists The acknowledgement that humans are not the evolutionary derived biological dispersal agents of the Gaian superorganism but have asserted their role to act as such exposes the Gaia theory as an active instrument in the legitimisation of human space expansion.

It might be pertinent to ask: 'who voted humans to be the reproductive representatives of the Earth?' Of course, humans did. The other members of Gaia had no say in the matter. According to the Gaia concept, however, it is easy to convince ourselves that it is right and proper to act in an imperialistic manner throughout the cosmos and then disguise such self-interest under a veil of 'just doing what comes naturally'.

Nigel Clark[67] has intimated that Gaia theory is part of a New Age philosophy that purports to be able to change the world by becoming personally more attached to it, much of the time through the continued development of advanced technology. Gaia's association with space development would tend to concur with this view. James Lovelock's Gaia is a technocentric Gaia which utilises high technology to advance her natural right to reproduce and colonise other worlds. Many environmentalist Gaians would certainly not like to be associated with such technocentric grandeur and arrogance but they might have trouble in avoiding it if they accept Lovelock's version of Gaianism.

The charge of grandeur and arrogance might be seen as solid by many readers but how can Gaia be criticized as imperialist given that there are no indigenous cultures in space? No indigenous cultures, may be, but indigenous life? That is another matter; especially with regard to the favourite candidate host planet for Gaian reproduction: Mars.

As talk of fossils and water on Mars permeates planetary science more and more, a revaluation of the life on Mars hypothesis is under way. If Gaians want to maintain their facade of non-anthropocentrism they have to admit that endangering the life of Martians (even if they are mere microbes) must in some way be ethically troublesome. The probability of the existence of native species upon Mars may be small but it is real and the effects of a Gaian expansion on such life-forms would probably be devastating.

Michael Allaby comments that in the 1970s "the Viking landers confirmed Mars to be lifeless"[68], and therefore there can be no objection to human interference with, and colonisation of, Mars. Margulis and Sagan agree:

> The Viking missions complemented ground-based astronomical observations and yielded definitive evidence for the lack of life on the red planet.[69]

Lovelock has agreed with this conclusion. Yet to declare that the Viking Landers did yield definitive evidence is hardly appropriate, however. Even within the scientific community this subject is heavily debated, especially within the context of recent scientific discoveries.[70] It is, of course, quite possible that Gaians tend to eliminate the ambiguities of the Viking results from their discussions because a negative result (regarding life on Mars) sits comfortably within their own scientific and metaphysical commitments. That the Viking Landers found no life is not clear for a lot of scientists researching the matter. All they can say is that it seems the Viking Landers were unable to find life in the few inches of dust that they tested at the two particular landing sites investigated. As Viking Project biochemist, Klaus Biemann, made clear with regard to the Viking results: "to say there is no life on Mars would not be a scientific conclusion"[71]

We may be able to forgive Lovelock's, Margulis' and Allaby's assertions of a lifeless Mars since, as Gaian advocates, they find it impossible to believe that extant life on Mars could exist because, according to the Gaian theory, life exists as a planetary-wide, global phenomenon. If there is life on Mars it must exist in an easily detectable and geographically extensive form that possesses robust planetary-scale homeostatic abilities.

Given their expansionist agenda, to have Mars colonised by Gaia's offspring, certain moral questions arise with regard to the faith of Gaians in their theory. It is of little consequence if scientists are willing to risk their professional credibility on the correctness of a theory suggesting life on Mars is impossible but for many environmental sympathisers, it would be considered intolerable arrogance that these scientists risk the extinction of extraterrestrial species on the basis of a yet to be proved theory which suggests that expansion to Mars can proceed with impunity.[72]

If, as the Gaians believe, life on any planet is only able to be present in a geographically extensive form and if it is the project of Gaians to give rise to the natural reproduction of Gaia then we must assume that their space exploration plans involve not a piecemeal 'flags and footsteps' approach to space endeavours but instead wholesale planetary-wide colonialism. The plan is not to land humans on Mars and scatter a few flags/experimental modules/laboratories about (as is, for instance, presently the case in Antarctica) but to colonise and engineer the planet with Earth life and human technologies in order to create a self-regulating biosphere like that of the Earth's. This is a "process known by tradition in science fiction as terraforming"[73] but which we are now encouraged to regard as Gaiaforming. Gaiaforming, Bonsor tells us, is:

> the controlled modification by scientific and technical means of a planetary environment, including its surface and, in particular its atmosphere, to establish a self-sustaining biological life-support system regulated by negative feedback loops.[74]

Lovelock and Allaby describe this process for a general audience in some detail within a fictional account of Martian colonisation entitled *The Greening of Mars*.

What is rather interesting in the writings of would-be terraformers and Gaiaformers such as Lovelock, Bonsor and Allaby is their particular conceptions about how life might develop. Plans for terraforming generally start with the seeding of lifeless planets with 'primitive' microscopic Earth organisms before more complex life-forms are introduced.[75] It is commonly perceived by terraformers that the development of life on a candidate planet from a microbial community to an advanced megaflora and megafauna community can be achieved by letting the processes of ecology run naturally along. In other words, terraformers and Gaiaformers are invoking the familiar ideas of Frederic Clements' succession theory whereby transplanted species will prepare the environment of other planets for the successful transplantation of more complex ecological communities. The science fiction writer Arthur C. Clarke adheres to this view in *The Snows of Olympus* by imagining the successive colonisation of a Martian landscape by lichens, then pines, then oak trees.

As discussed in Chapter Two, however, the dubious validity of succession theory might cast doubt upon the efficacy of any human sponsored program of ecological succession on this planet let alone one to which terrestrial organisms are not use to. Because succession theory has been more or less slain, terraforming ecologists may be left with alternative models of ecological change that suggest that succession from simple pioneering species to complex ecosystems is unlikely. Despite supplying a continuous rain of seeds of the desired species, terraformers would probably find themselves unable to develop their chosen ideal ecosystem. Instead of creating planets filled with enchanting forests and lakes reminiscent of the Earth's great terrestrial wildlands, humans may create nothing but millions of square kilometres of pungent microbial bogs more redolent of Earth's polluted industrial waste sites.

The creation of new Earths in the vein of Gaiaforming might be used to make out that space exploration is an environmental activity. People like Martin-Smith and Arthur C. Clarke pretend that it is the ultimate in environmental activities since it involves the creation of new living planets instead of their destruction, as seems to be occurring with the Earth, yet for Gaia's space children there is another way in which space exploration is intimately environmental. Like many others who are fascinated by space exploration, James Lovelock, Michael Allaby, Carl Sagan, Dorian Sagan and Lynn Margulis champion space development as an ecological endeavour since it gives rise to new ways of comprehending the Earth's unity. Indeed for them, and for many others, the unity of nature can best be understood and appreciated via the space photographs/images of the Earth from space. The following is a typical example of the use of such Earth imagery:

> Observing the Earth silhouetted against the black backdrop of space, one can hardly deny its wholeness and fragility. Earth System Science was born out of the Space Age and is a holistic approach to studying the Earth as a system of interacting components.[76]

However, it might be said that the Earth image does not only symbolise the beauty, fragility and unity of the world we inhabit. It also symbolises the feats of technology that industrialism has offered that world. Space enthusiasts continue to explicate the environmentalist nature of space exploration but the irony of the Earth image escapes them. Only by conquering the terrestrial environment and making nature the dutiful servant of technology, has humanity managed to propel itself spacewards. The space photographs that capture the Earth image stand as testament to this all-conquering victory over the natural world. Or as the philosopher, Michael Zimmerman, sees it:

> In such photos, we see Earth reflected in the rear-view mirror of the spaceship taking us far away from home in order to conquer the universe.[77]

Conclusion: Gaia as an Environmental Goddess?

If James Lovelock has his way the appreciation of the unity of nature would manifest itself in a global systems view of the world. Lovelock says that if such a view was explored scientifically to its completion we would be able to locate and identify all the various physically interconnected pathways that exist in nature so that we do not disturb them. According to Lovelock, and his many environmental supporters, if such a view was explored philosophically to its completion we would appreciate the integration and interdependency of the natural world and come to realise the organic nature of all things. This chapter has attempted to show that even if we do this, even if we view the world as an organic, interdependent, global Gaian system, the environmental crisis will still continue since the unity spoken about within Gaian theory fails to capture the values of the living members of the Earth that are being overwhelmed by that crisis. The unity idea, especially as manifested through Lovelock's Gaia, would also mislocate the causes of environmental destruction in 'natural' processes such as over-population. What these faults, taken together, might suggest is that the Gaia theory is hardly in a position to offer sound environmental narratives about the world. Not only that, however, Gaia theory legitimises and advocates a program of space imperialism which might also endanger the environments of other worlds. Gaia, it must be concluded, is not an environmental Goddess but a technocentric Goddess, one that tolerates rampant industrialization and extinction in all but her vital zones, and one that cares little for the individual biotic members of this planet (and others).

Notes to Chapter 3 (Gaia)

1. Indeed this is the reading offered of Gaia in the following texts: D. Abram, "The Perceptual Implications of Gaia, *The Ecologist*, 15, no. 3, (1985): 96-103; M. Allaby, *Guide to Gaia*, (Optima Books: 1989); P. Bunyard and E. Goldsmith, eds, *Gaia and Evolution*, (UK: AbbeyBodmin, 1989); L.E. Joseph, *Gaia: The Growth of an Idea*, (NY: Arkana, 1990); J. Lovelock, *Healing Gaia: Practical Medicine for the Planet* (NY: Harmony, 1991); T. Roszak, *The Voice of the Earth*, (NY: Simon and Schuster, 1991); S. Nicholson and B. Rosen, eds, *Gaia's Hidden Life:* (Wheaton, Ill: Quest Books, 1992);

E. Goldsmith, *The Way: An Ecological Worldview* (NY: Shambhala, 1993); C. Merchant, ed, *Key Concepts in Critical Theory: Ecology* (NJ: Humanities Press, 1994); F. Capra, *The Web of Life*, (London: HarperCollins, 1996); T. Volk, *Gaia's Body: Toward a Physiology of the Earth* (NY: Copernicus, 1998).

2. R. Lewin, "All for One, One for All" *New Scientist*, 152 (14 Dec 1996), p29.
3. J. Bonsor, "Gaiaforming", *Amateur Astronomy & Earth Sciences*, 2, no. 2, (1997): 26.
4. See, for example, Bunyard and Goldsmith, eds, *Gaia and Evolution*; Joseph, *Gaia: The Growth of an Idea*; and Lewin, "All for One, One for All".
5. Goldsmith, *The Way*, p.xviii.
6. See: Abram, "The Perceptual Implications of Gaia"; Allaby, *The Guide to Gaia*, Roszak, *The Voice of the Earth*; Merchant, *Key Concepts in Critical Theory: Ecology*.
7. See, respectively: Goldsmith, *The Way*, p.11, and Goldsmith, *The Way*, p.105.
8. See, for example, claims of non-anthropocentrism for Gaia in Goldsmith, *The Way*, as well as in J. Lovelock *Gaia: A New Look at Life*, (Oxford: Oxford University Press, 1979) and J. Lovelock, *The Ages of Gaia* (NY: WW. Norton & Co, 1988).
9. A. Brennan, *Thinking About Nature* (Athens: University of Georgia Press, 1988), p.131.
10. Lovelock, *The Ages of Gaia*, p.11.
11. Lovelock, *Healing Gaia*, p.9
12. Lovelock's disciple, Tyler Volk, constructs an analogy of the Gaian organism by using the human body:

> "just as the organs occupy unique locations within us, so the biomes spread across the planet in distinct biogeographical provinces: tropical rainforests, savannas, deserts, temperate grasslands, temperate forests, boreal forests, tundra. This dissection of Gaia into parts has immediate appeal because the viscera of Gaia would then be visual biological regions." From T. Volk, *Gaia's Body*, p95.

13. Lovelock, *The Ages of Gaia*, p.xvii.
14. J. Lovelock, "A Model for Planetary and Cellular Dynamics", in W.I. Thompson, ed, *Gaia: A Way of Knowing* (MA: Lindisfarne Press, 1987) p127. The same sort of argument can be found in S.A. Levin, *Fragile Dominion: Complexity and the Commons* (Helix Books, 1999).
15. Lovelock, *The Ages of Gaia*, p.179.
16. J. Lovelock, "Gaia", in C. Merchant, ed, *Key Concepts in Critical Theory: Ecology*, (NJ: Humanities Press, 1994) p.357.
17. Lewin, "All for One, One for All", p.33.
18. Brennan, *Thinking About Nature*, p.191
19. The two obvious problems with the tyranny of the niche are: 1) it is always going to be impossible for human science to work out all the various jobs of every single one of the world's biotic members. So if we end up valuing species for nothing else but their jobs then we are likely to ignore and devalue all those invisible and unseen jobs which are performed in the ecological world in unnoticed ways or by unnoticed species. 2) those

species that do not have a job, or do not appear to have a job, are immediately regarded as valueless.

20. Lovelock, *Healing Gaia*, p16.
21. R. Carson, *Silent Spring* (Boston: Houghton-Mifflin, 1962). Carson is generally regarded as one of the founders of the modern environmental movement because she informed a wide audience about the disastrous affects of chemical pollution in the biotic environment.
22. Lovelock, *The Ages of Gaia*, p.219.
23. L. Westra, *An Environmental Proposal for Ethics*, (Baltimore: Rowman and Littlefield, 1994) p.xi.
24. L. Margulis and D. Sagan, "Can Mars be Colonized?", *Paper Presented to the 29th Plenary Meeting of the International Astronautical Federation*, Washington, D.C., 1992), p.1.
25. Volk, Gaia's Body, p23.
26. Brennan, *Thinking About Nature*, p.191.
27. Volk, Gaia's Body, 127.
28. A. Gare, *Postmodernism and the Environmental Crisis* (London: Routledge, 1995) p.100.
29. D. Pepper, *The Roots of Modern Environmentalism*, (Beckenham: Croon Helm, 1984), p.241. There is a case to be made that technocentrism, especially Lovelock's technocentrism, should not be classed as a type of environmentalism at all since it is only very loosely worried about environmental problems and the need find solutions tò them.
30. For instance, see Pepper, *The Roots of Modern Environmentalism*, and B. Devall and G. Sessions, *Deep Ecology: Living As if the Earth Really Mattered* (Layton: Gibbs M. Smith, 1985).
31. Joseph, Gaia: *The Growth of an Idea*, p.245.
32. Lovelock, *Healing Gaia*, p.155.
33. *ibid.*
34. See M. Cromartie, ed, *The 9 Lives of Population Control*, (Grand Rapids: Eerdmans, 1995); A. Durning, *Misplaced Blame: the Real Roots of Population Growth* (Seattle: Northwest Environment Watch, 1997); C.R. Holm, eds, *Population: Opposing Viewpoints*, (San Diego. Greenhaven, 1997)
35. Lovelock, *The Ages of Gaia*, p.179.
36. Lovelock, *Healing Gaia*, p.155.
37. *ibid.*
38. Brennan, *Thinking About Nature*, p.131.
39. F.B. Golley, *A History of the Ecosystem Concept in Ecology*, New Haven: Yale University Press, 1993) p.58.
40. J. Hagen, *An Entangled Bank: The Origins of the Ecosystem Concept* (New Brunswick: Rutgers University Press, 1992), p191.
41. Lovelock, *Healing Gaia*, p.27.
42. See S.E. Kingsland, *Modeling Nature*, (Chicago: Chicago University Press, 1985) and S.E. Kingsland, S.E. "Economics and Evolution: Alfred James Lotka and the Economy of Nature", in P. Mirowski, ed, *Natural Images in Economic Thought* (Cambridge: Cambridge University Press, 1994) p.231-248.
43. Volk, *Gaia's Body*, p.xiii-xiv.

44. Lovelock, *Healing Gaia*, p.20
45. *ibid*, p.6.
46. Lovelock, *The Ages of Gaia*, p.xvii.
47. L. Margulis quoted in Joseph, *Gaia: The Growth of an Idea*, p.70.
48. The Daisy World model served as an answer to the scientific criticism that came from Richard Dawkins and W. Ford Doolittle who criticized the Gaia theory on the grounds "that there is no way that diverse living organisms of the Earth could act in symbiosis...to regulate the planetary environment" From Lovelock, *Healing Gaia*, p.62.
49. Lovelock, *Healing Gaia*, p.7.
50. It might be contended by some readers that my delineation in this section between 'science' and 'philosophy' is artificial or arbitrary. In other words, what Lovelock is doing--what any other scientist is doing--is both science *and* philosophy. This view would contend that in the course of their professional life, scientists must necessarily be philosophical, i.e. engage in debate with non scientific ideas, metaphorical concepts, and broader populist worldviews, since they are not just engaged in unraveling the truth of the real world but are also involved in interpreting it for others. I'm sympathetic to such a view and it should be noted that these issues are to be addressed in later chapters.
51. See, respectively: Lovelock, *Healing Gaia,* p.16. and p.158.
52. See, for instance, B. Devall and G. Sessions, *Deep Ecology: Living As if the Earth Really Mattered*; Brennan, *Thinking About Nature*, as well as: J.B. Callicott, *In Defence of the Land Ethic: Essays in Environmental Philosophy*, Albany: SUNY Press, 1989); A. Naess, *Ecology, Community, Lifestyle* (Cambridge, U.K., Cambridge University Press, 1989); M. Bookchin, *The Philosophy of Social Ecology: Essays in Dialectical Naturalism* (Montreal: Black Rose, 1990); S.J. Armstrong and C.J. Botzler, eds, *Environmental Ethics: Divergence and Convergence* (NY: McGraw-Hill, 1993); M.E. Zimmerman, *Contesting Earth's Future: Radical Ecology and PostModernity*, (Berkeley: University of California Press, 1994); R. Elliot, ed, *Environmental Ethics*, Oxford: Oxford University Press, 1995).
53. Lovelock, *Healing Gaia*, p.158
54. Lovelock, *Healing Gaia*, p.16.
55. D. Sagan and L. Margulis, (1988) "Gaia and Biospheres", in P. Bunyard and E. Goldsmith, eds, *Gaia: the Thesis, the Mechanisms, the Implications*, (Wadebridge, UK: Quintrell, 1988) and L. Margulis and G. Hinkle, "The Biota of Gaia", 150 Years of Support for the Environmental Science", in S.H. Schneider and P.J. Boston, eds, *Scientists on Gaia*, Cambridge: MIT Press, 1991) p.11-18.
56. See Lovelock and Allaby, *The Greening of Mars*. (NY: Warner, 1985).
57. L. Margulis, L. and D. Sagan, "Can Mars be Colonized?".
58. *ibid*, and also see: L. Margulis, and O. West, "Gaia and the Colonization of Mars," *GSA Today*, 3 (no. 11), p.277-291 (1993).
59. R. Schweickart, "Gaia, Evolution, and the Significance of Gaia", *IS Journal*, 2, no. 2, (1987) p.29.
60. Lovelock, *Healing Gaia*, p.21.
61. *ibid*.

62. Which should not be confused with the ecological equilibrium of ecosystem ecologists which is a different phenomena altogether.
63. Margulis and Sagan, "Can Mars be Colonized?"
64. M. Martin-Smith, Microcosm and Macrocosm, *Quest*, 1, 3, (1997) p.31.
65. Schwieckart quoted in Joseph, *Gaia: The Growth of an Idea*. p.63.
66. Allaby, *Guide to Gaia*.
67. N. Clark, "The Re-Enchantment of Nature? The Politics of New Ageism and Deep Ecology", in *Ecopolitics III: Proc. of the 3rd Ecopolitics Conference*, Hamilton, NZ (1988).
68. Allaby, *Guide to Gaia*.
69. Margulis and Sagan, "Can Mars be Colonized?", p.1.
70. See: B.E. Digregario *et al*, *Mars: The Living Planet*, (Frog Limited, 1997); D. Goldsmith, *The Hunt for Life on Mars* (NY: Plume, 1998); L. A. Leshin "Insights into Martian Water Reservoirs from Analyses of Martian Meteorite QUE94201", *Geophysical Research Letters*, 27, (2000) No:14; M.C. Malin and K.S. Edgett, "Evidence for Recent Groundwater Seepage and Surface Runoff on Mars" *Science* (2000) :2330-2335.
71. K. Biemann, "Search for Organics", in *Viking Mars Expedition*, 1976, MMR PR Dept, CO, (1978) p.32.
72. Both Margulis and Lovelock continually talk about speaking for, representing, and defending microbes. Lovelock says that he and Margulis are not directly interested in the fate of humans but instead are interested in the "much maligned microbes with which the Gaian system originated and which continue to do its basic work" (quote from Lovelock, *Healing Gaia*, p.101.) This is somewhat ironic, however, if we consider how they adopt an attitude that positively endangers the lives of whole species of microbes on other planets by promoting the invasion by murderous hordes of Earth bacteria. See, for instance: A. Marshall, "Ethics and the Extraterrestrial Environment" *Journal of Applied Philosophy,* 10, 227-237 (1993); A. Marshall, "Gaian Ecology and Environmentalism", *Wild Earth*, 7 (1997) 76-81.
73. Bonsor, "Gaiaforming", p.27
74. *ibid*, p.26.
75. See for instance: J. Lovelock and M. Allaby, *The Greening of Mars*, (NY: Warner Books, 1985); C. Sagan, Cosmos, London: MacMillan, 1981); A.C. Clarke, *The Snows of Olympus*, (London: Victor Gollancz, 1996).
76. This quote comes from the following website:

    http://web.geology.ufl.edu/earth6.html.

    Writings which posit the environmental nature of space exploration by utilising similar Earth imagery include: Sagan, *Cosmos*; Capra, *The Web of Life*; Martin-Smith, "Microcosm and Macrocosm".
77. Zimmerman, *Contesting Earth's Future*, p.75.

CHAPTER 4

Natural Conservatism:
The Unity of Nature and Social Systems.

Introduction: Unity Between Ecology and Society

If we claim that unity as manifested through systems ecology and Gaia theory is holistically over-zealous and if it is acknowledged, by unitarians and others, that there is unity between ecology and society, between nature and humanity, then might not the same sort of critique of over-zealous holist tendencies apply in the social realm? The obvious thing to want to find out when addressing such a problem is to ask: 'how exactly do unitarians see the unity relationship between ecology and society, between humanity and nature?' This question is important since there may be different ways that humanity and nature can be said to be united. Some of them benign, some of them not.

Some environmentalists, like Edward Goldsmith, see the humanity-nature relationship very strongly. Goldsmith would say that there are values in the biosphere, that ecological knowledge reflects those values, and that those values must be adopted in order for both humanity and nature to survive as we would like.[1]

One of the most common ways of identifying nature with society is by classing them both in economic terms. Callicott, an environmental philospher, does this for instance when he says:

> The various parts of the 'biotic community' (individual animals and plants) depend on one another economically so that the system as such acquires characteristics of its own. Just as it is possible to characterise and define collectively peasant societies, agrarian communities, industrial complex, capitalist, communist, and socialist economic systems and so on, ecology characterises and defines various

> biomes as desert, savannah, wetland and tundra, woodland and other communities, each with its particular 'professions', or 'niches.'[2]

Here again we see, even if only in an over-generalising example, the issue of functionalism emerging since it is perceived by Callicott that functions are analogous across the humanity-nature boundary. Robert Ayres, in another way, has also analogised about the similarities between the economic and biological realms in terms of functions:

> For an economic system, the analogues of cells are individuals, the analogues of specialised organs are firms and industries; and the products of industry are analogous to the minerals, amino acids, sugars, and lipids as well as the more specialised vitamins that circulate through the organisms and are metabolised for energy and/or incorporated into biomass.[3]

Similarly, as Golley points out, Eugene Odum, the greatest campaigner for ecosystem ecology in the 1950s and 60s, was never afraid to analogise with economic references, interchanging–as he did--money for energy and jobs for niches when he attempted to diagrammatically represent the ecology of certain regions.[4]

Unity via the 'System'

Ayres, like other philosophers of nature who see humanity and nature as united entities, has often provided exact explanations of the basic ways that human and natural systems intersect:

> An economic system is, in three important respects, a kind of living system...firstly...an economic system is similar to biological systems in exhibiting self-organisation...secondly...[both have] the need for continuous inputs of useful energy...thirdly...[a] strong similarity between economic systems and living systems is their ability to grow and evolve.[5]

We can see in Ayres' passage the ease to which simple analogies become complex when filtered through the lens of systems theory. Instead of comparing 'nature' and 'economics' as Callicott does, Ayres compares 'living' and 'economic' systems. Systems theory enables similarities to become far more transparent. Once the systemic nature of something or other is identified, it is very easy to find the way it resembles the properties of completely different things which have also been systematised.

This description of quite different things as 'systems' and then their subsequent comparison, is an avowed goal of systems science and always has been. The search for parallel principles in diverse phenomena is neatly summed up, with the help of a neologism, by one of the progenitors of systems science; Ludwig von Bertalanffy, who describes the presence of similar principles in different areas of study as 'isomorphy'. "The search for such isomorphies is a major pursuit of systems science." Bertalanffy says.[6] He then goes on to summarise what isomorphy is:

> isomorphy... is a consequence of the fact that in certain aspects, corresponding abstractions and conceptual models can be applied to different phenomena. It is in view of these aspects that system laws will apply.[7]

Speaking about Bertalanffy from a historical point of view, Hagen summarises his program of systems science and cybernetics like this:

> cyberneticians were attempting to construct an overarching mathematical theory to explain the behaviour of organisms, machines and other complex systems. In fact, from the perspective of cybernetic analysis, distinctions between machines, organisms and even societies seemed to evaporate; all were treated mathematically as 'systems'.[8]

In 1967, Bertalanffy made friendly warnings about the supposed closeness of systems analogies. Just because analogies are made between certain conceptual models:

> this does not mean that physical systems, organisms and societies are all the same. In principle, it is the same as the situation as when the law of gravitation applies to Newton's apple, the planetary system, and the phenomenon of tides. This means that in view of some rather limited aspects a certain theoretical system, that of mechanics holds true; it does not mean that there is also a particular resemblance between apples, planets and oceans in a great many other respects.[9]

It is interesting that Bertalanffy has to explain the relationship of systems analogies by using an analogy but his warning nevertheless seems to be that systems analogies are functional, not structural. The apple, the planets and the oceans are structurally dissimilar but they function in relation to gravity in the same way.

However, maybe it's wise to be sceptical of Bertalanffy's assurance that function is all that matters in systems science as it tries to find isomorphies (and to be skeptical, also, that there are no supposed structural similarities between the components of one system and the next). At least in the abstract, systems scientists are not adverse to assuming that systems across different disciplines/subjects/worlds are in fact as similar in structure as they are in function. One way they do this is through the use of the concept of hierarchically arranged levels. For example, Ayres tells us that there is:

> the obvious and close analogy between an economic system and the food web of an ecological system or community. In the latter case one can consider each species as a node. The energy inputs to any node are derived from other species from a lower trophic level, except in the case of the lowest level (vegetation), which derives its energy from photosynthesis. The node and flow structure can be defined in diagrammatic form (and when we do so we find) not surprisingly, essentially the same diagram applies to the economic case.[10]

Ayres also intimates that the economic and ecological systems that exist on the dipole of his analogy are so close that "the corresponding mathematics is almost identical if one conceptually substitutes money flow for energy flow".[11]

When Ayres talks about the identical structure and functions of various processes in ecology and economics he might be considered to be finding isomorphies; actual existing objective similarities between one worldly frame of reference and another. This was Bertalanffy's plan; not to create analogies based on metaphors of similarity taken from one tradition to another but to discover real similarities.

In contrast to Bertalanffy, Callicott seems to realise that his analogies are metaphorical since when Callicott talks about the identical structures of economic and ecological settings he knows he is simply taking the language of one discipline and applying it to another. From Callicott's point of view this does not make his own economy/ecology associations weak since his use of a particular metaphor assumes that there is sufficient resemblance between economics and ecology for the metaphor to work in a sustained way. While Callicott can accept that he is being metaphorical, Ayres--via systems science--is convinced that the comparisons are reflections of actual isomorphies.

Whether or not economic/ecological metaphor users admit to their comparisons being metaphorical or not, it often seems that in situations such as this the appropriate thing to do is to identify the site from where the metaphor comes and having found it, expose it as mere metaphor and thus undermine its truth. If we abide by this idea (which has its problems[12]) we then might go on to ask where the original referent lies in the ecology/economics analogy.

If ecology is interpreted as being fundamentally the same as economics by Callicott, and alternatively, if economics can be fundamentally based on ecological principles, as for instance Ayres sometimes says, then which one is the original referent? Which one should we expose as the masked metaphor that must be revealed as such? Was it ecology that supplied economics with a grounded base from which to analogise from or vice versa. The answer seems to be that it is impossible to state which subject served as the referent and which one was being the metaphor. It is not that ecology gave Ayres economical principles or that economic principles gave Callicott the principles of ecology. Both economics and ecology have had such a long and tortuous history of cross-fertilisation that principles within one feed from the other and then feed back to the 'original'. We may call these types of narratives 'reprojective spiral narratives', a term that Phillip Mirowski[13] is driven to use when examining the nuances of the history of natural images in economics.

The relevance of these reprojective spiral narratives to this book is in directing us away from just trying to stick to an approach of identifying from whence the original metaphors came so as to expose them as mere metaphors. For instance, when trying to work out why nature is considered in unity with society we have to explore the use of metaphors taken from social science and natural science that say it is so but I suggest that we will get nowhere if we permanently look for the first pure original metaphor because both social science and the natural sciences have been feeding off of each other for so long.[14] Following the twisting past of metaphors, however, might open up some of the intellectual heritage of the unity of nature idea (and this is what is attempted in this chapter and also in chapter seven).

Unity and Social Holism

If humanity and nature are united so that they obey at least some of the same rules, as the systems thinkers suggest, then what sort of society do these rules create? Might such rules inflict the same problems on society as they do on united and systematised ecological settings? Will they suffer from being functionalist and thereby ignore the individual? Will they suffer from a lack of acknowledgement of change and thus form a conservatism of a social type? These questions have been posed by numerous environmental scholars and it is their arguments which form the basis of the next few sections of this chapter.

One of the most obvious points to note about unitarian ecological theory is its closeness of approach to the sociological tradition known as 'structural functionalism'. Both systems ecology and structural functionalism draw on a common theoretical affinity involving holism and systems theory.

The orthodox way for recounting the philosophy of structural functionalism is by saying something like:

> Functionalism uses an organismic model of systems as the basic reality, draws on concepts of structure and function from physiology and stresses adjustment or adaptation as a goal for organisms, social groups or whatever.[15]

Of these migrations of biology into sociology, Tony Benton writes:

> Overwhelmingly, biological intrusions into the domain of the human sciences have been politically conservative, and this is especially true of their widespread popular diffusion.[16]

However, it may not be so much of a one-way biological intrusion as it is a case of two way traffic. Organicism in nature supports organicism in sociology and vice versa. Sometimes it is impossible to tell whether the original application was one way or the other. Even in the writings of particular scholars it is not clear whether the terminology is taken to be primarily social or natural and then applied or reapplied from one to the other.[17]

Although social functionalists are dedicated to the study of social situations, in many cases distinct analogies/metaphors are drawn from the biological world. At other times, analogies to the biological world are avoided by social functionalists but references to the works of influential scholars who themselves talk about the simultaneous biological/social applicability of their work is evident. Thus a structural functionalist social scientist may be able to superficially de-link his or her work from organicism and biology (and ecology) but this does not negate the presence of both philosophical and historical linkages.

In some cases, the association is more distant (to the point that some structural functionalists and some ecosystem scholars have probably never cited a writer relevant to both). Talcott Parsons, the most famous advocate of structural functionalism, for instance, never attributed any of his social systems thinking to the systems thinking of those that have become known as the principal 'fathers' of systems theory (such as Ludwig von Bertalanffy, Norbert Wiener and John von Neumann) whereas Eugene Odum, Robert Ayres and Fritjof Capra make this attribution for their own work quite enthusiastically. However, the same people that influenced Parsons, people like Lawrence Henderson and Walter Cannon, are known to have influenced Bertalanffy, Wiener and von Neumann.[18]

Despite the sometimes tenuous links, it can be claimed, as Lilienfeld does, that the ideas of biological systems fans and social systems fans are so strongly alike that they may be placed upon the same 'analytic operating table':

> Parson's 'system' may be classified as belonging to systems thinking; certainly its philosophical categories and its metaphysics are the same.[19]

If this is correct, any critique of the unity of nature idea and of ecosystems philosophy must also make forays into social systems theory.

Where ecosystem ecologists honour the ecosystem as the basic unit in ecological study, so structural functionalists honour the social system as the basic unit in sociological study. Let us see how these two approaches draw from these common thematic premises and so suffer from common problems.

Social Systems and Change

Just like those who work with ecosystems, so those who work with social systems tend to emphasise stability over change: "any system that does not have some ability to resist perturbation tends to cease to exist as a system."[20] Within social science this metaphysical or methodological commitment to stability has often lead to systems-inclined sociologists being labeled as politically conservative by non-systems-inclined sociologists since the former variety of sociologist are involved in legitimising the predominating social order. This is most evident in the case of the structural functionalist tradition of sociology which was, during the mid-Twentieth century, the dominant Western sociological tradition.

The primary proponent for this tradition was Talcott Parsons. For many sociologists writing during the decline of structural functionalism in the 1960s, Parsonianism was thought of as "a celebration and affirmation of the status quo."[21] Today a non-functionalist sociologist might sum up the principle ideas of structural functionalism in this way:

> Parsons is essentially a conservative thinker who ignores the exploitative and unequal character of capitalism, along with the resulting division and disharmony.[22]

If we place this sort of criticism within its historical context it becomes clear just what sort of status quo, and what sort of stability Parsons, and most of the other structural functionalists, were aiming for:

> it is important to understand Parsons work as an attempt to defend capitalist society against the criticism contained in Marxist analysis. Although both Marx and Parsons see capitalism as a 'social system', their assessments of it are very different. Marx envisioned capitalist society as basically exploitative (of the working classes), conflict-ridden, and governed primarily by the profit motive inherent in the economic system. On the other hand, Parsons recognised that capitalism was still striving towards its ideal form, he saw it as a basically fair and meritocratic system.[23]

Marxists might be keen to add to this conceptualisation of Marxism that:

> with its emphasis on dialectics, permanent upheaval and irresolvable conflicts between classes, Marxism would seem incapable of adopting an organismic view of society, one which is in essential balance.[24]

Although, historically, structural functionalism was a way to show how mid-Twentieth century U.S. capitalist society was a wonderful, freely self-adjusting system that was incapable of being perturbed greatly from within, the concept of the social system that Parsons advanced can actually be used to show how any social system might operate. Edward Goldsmith, for instance, uses Parsonian sociology during his sustained and enthusiastic celebration of traditional (premodern and precapitalist) societies, which are to him, the only truly Parsons-like self-regulating societies that exist. Just as change in the social system for many, if not most, structural functionalists is something either strange or undesirable, so social change, for Goldsmith, is highly undesirable.

If, as Goldsmith would like us to believe, "stability rather than change is the basic feature of the living world"[25] (and if human communities and ecological communities are of the same structure) then stability is the basic feature in the human world too. This is the line that Goldsmith takes in *The Way*; he merely maintains that modern societies do not adhere to such natural stability while traditional societies do. So what do people in traditional societies do in order to operate in a way that is as stable as ecosystems? Well, basically, they do the same things that Parsons said people do to keep the American social system lovely and stable:

-the members of the community have a common belief system and common values and those that do not adhere to these values are ejected or re-educated and rehabilitated.

-the members fulfill their role within the community as per these common beliefs and values. Those that do not are ejected from the system or re-educated and rehabilitated.

Because Goldsmith presents his traditional societies as operating in this Parsons-like way, they may be adjudged as being incapable of tolerating dissent and differences of opinion and lifestyle. Such a judgment would fall heavily upon Parson's 1950s America and it may fall heavily on Goldsmith's traditional societies too.

Although their ideal societies are very different, the Parsons and Goldsmith variants of stable societies have much in common. Parsons emphasised common values, as does Goldsmith. Parsons emphasised the essential role of the family, as does Goldsmith. And Parsons emphasised tightness of tradition, as does Goldsmith.

Parsons, like Goldsmith theorizes today, also had a hierarchy of systems within systems that resembled the levels of organisation posited by modern day proponents of systems theory. Just like the levels of biological organisation presupposed by systems ecologists, Parsons' levels were structured in the same manner. Where

Goldsmith (as well as Capra and Davies and Lovelock) would present the interaction between cellular systems, organismal systems, ecosystems and the global Gaian system so that the lower levels contribute to the higher levels which then act to extend influence back upon lower levels, so Parsons invented a four-level system of systems in which the body system, the psychology system, the social system and the cultural system were all nestled within each other so that lower systems gave rise to higher systems which in turn extended influences on the lower systems once more.

So if the principles that run, stabilise and harmonise Goldsmith's and Parsons' social systems are the same, how can they give rise to such varying social visions; in Parsons' case, mid-Twentieth century Western capitalist-industrialism; and in Goldsmith's case ecologically-benevolent (neo-)traditional societies? This is a relatively unimportant question in this book since it is by no means proven that Parsons' social systems were operating to produce his beloved America or that Goldsmith's social systems can possibly operate in traditional societies to give rise to his beloved eco-friendly communities. What is of interest is the way that they both use changelessness and stability--and hence intrinsically conservative regimes--to legitimise the societies they variously desire.

Goldsmith does not have much time for Parsons' affirmation of capitalism but he does share Parsons' love of systems. The great difference between the two, so says Goldsmith, is that modern American society is not stable while traditional (precapitalist) society is. Because traditional societies are living in stability within their own ecological setting and so are not damaging either Gaia or the ecosystems that make her up, says Goldsmith, then it is toward the direction of traditional societies that modern society should aim.

According to Dickens[26], Parsons saw traditional societies not as ecologically-friendly and socially-benevolent (as Goldsmith does) but as primitive, undeveloped and undifferentiated societies subject to perturbation. Modern systems, on the other hand, were, to Parsons, classifiable as complex and self-regulating. This varying evaluation of traditionality exemplifies two extreme approaches to pre-modern societies. Traditional societies were noble and enlightened according to Goldsmith, while being savage and uncivilised according to Parsons. Both versions sit within the extremes of western myth-making. Whether a social thinker is prejudiced toward the view that all traditional societies were harmonious with their environment and a joy to live in or prejudiced toward thinking they were all savage, undifferentiated and undeveloped societies, these views by themselves tend to be hopelessly homogenous views of vastly-differing societies (that are actually incredibly diverse, even if only along a single characteristic of interest such as ecological-friendliness or internal complexity). Some traditional societies may be ecological friendly and socially-benevolent as Goldsmith suggests, but not all. And some traditional societies may be internally simple and undifferentiated as Parsons suggests but, again, not all.

Social Systems and Functions.

The major relationship between Goldsmith's traditional societies and Parsons' 1950s American society is in their parallel appreciation of the relationship of the whole society to their respective individual parts. Just as Parsons' social parts (individuals, families, institutions etc.) were arranged to contribute to the stability of the whole society, so Goldsmith picks his social theories straight from Parsonianism to suppose that traditional societies are arranged so that the parts contribute to maintaining the whole. Here, the parts are, like the parts in an ecosystem, ascribed a role or a function. Individuals, families and institutions are occupiers of a particular social niche which is implanted within an interdependent whole via common values in order to maintain the entire society. This emphasis on society being a functional system is common to both Parsons and Goldsmith. Philosopher Jean-Francois Lyotard would have explained this obsession with social niches like this:

> the decision makers attempt to manage these clouds of sociality according to input/output matrices, following a logic which implies that their elements are commensurable and that the whole is determinable.[27]

Whilst sociologist Zygmunt Bauman might explain it like this:

> Constantly lurking behind the scene in the orthodox vision of social reality was the powerful image of the social system--this synonym of an ordered, structured space of interaction, in which probable actions have been, so to speak, pre-selected by the mechanisms of domination and value sharing. It was a '...co-ordinated space...', one inside of which the cultural, the political and the economic levels of supra-individual were all reservant with each other and functionally complimentary.[28]

My point here is that these same descriptions are applicable not only to Parsons (where they were aimed) but to Goldsmith as well. Goldsmith's social function ideas posit that traditional societies have co-ordinated parts that are complimentary in their functionality so that an ordered whole is the result.

One of the functional parts common to both Goldsmith and Parsons is the family. Goldsmith, writing in 1993, approves of 'traditional' families and looks distastefully on 'modern' single-parent families since:

> a single parent family [is] a highly unstable entity that does not provide a satisfactory environment for the children's upbringing, and which is likely to break

> down further, leading to abandonment of the children as is happening on such a scale in the slums of South America.[29]

For both Goldsmith and Parsons, families are an essential level in the make up of social systems. They enable education, reproduction, value installment and the appropriate partitioning of resources within society. For both of them the death of the family would be the death of the social system (and thereby the death of the ultimate system; Gaia in Goldsmith's case, and America in Parson's case). However, Goldsmith's restrictive definition of what it is to be a family would strike many as being highly exclusionist. If Goldsmith's eco-society eventuates it may be just as narrow and morally conservative as Parson's 1950s America where the predominant values excluded many who did not come up to Parson's vision of what it took to be functional family.

Social Systems and the Individual

Social systems and the organic unities that they are said to reflect, have sometimes been criticized for being anti-individualistic. We might, for instance, note the following description of an organic view of nature:

> Connectedness involves regulatory feedback that controls component dynamics to meet the best needs of the organisation itself, not necessarily the best needs of the components (e.g. programmed cell death during embryogenesis or ejection of drones from beehives are detrimental to the components but benefit the organisation as a whole).[30]

If an organic view of nature is to be read like this, then might not an organic view of society based upon such a view of nature be anti-individual? This is a worry that has permeated environmental philosophy for all of its history and some environmental holists have grown annoyed at it. According to the holists, any new socially-aware anti-unitarian critiques of environmentally-guided holist ideas:

> are only the latest in a long line of indignant but vastly predictable accusations that continue to be reacted monotonously. Detailed responses to these criticism have been offered for years.[31]

This extract from Laura Westra goes on to inform us that although some major thinkers have provided carefully argued counter-claims, it appears that "no argument so far has succeeded in silencing the critics" and their "lack of open-mindedness or interest."[32]

We can interpret Westra charitably to mean that she admits that holists have trouble getting their point across to those who choose to remain ignorant or we can choose to interpret Westra's statement uncharitably to mean that she wishes to close down the ongoing debate about the connections between holism, conservatism and anti-individualism. In this section and the next, the debate as conducted by previous authors is touched upon so that it might be used and developed in later chapters.

The arguments of the "anti-holists" declare that, at least in the abstract, individuals are relegated to secondary importance compared to the value of the whole. Despite Westra and others' protests that environmental holism does not lead to social holism there are explicit examples within environmental thought that seem to show that it is quite possible. At the forefront is Edward Goldsmith who declares that: "To keep to the Way, society must be able to correct any divergence from it."[33] The Way, Goldsmith makes clear, is the specific and general rules that run and maintain the holist entity that is Gaia. These rules are, at once, social and biological.

Divergence from these rules can be normally handled, says Goldsmith, via the various self-regulating processes within the various systems of Gaia, be that system of a social nature or of an ecological nature:

> If the Gaian hierarchy is to maintain its stability, all individual living things that compose it must obey a veritable hierarchy of laws, which together constitute the laws of nature.[34]

These rules thus impose some sort of boundaries, parameters and self-discipline upon the individual. Both humans and non-humans must live in a way commensurate not only with their immediate social and ecological system but with the greatest system of all; Gaia. In stronger words, this might be translated to mean that the freedoms, lifestyles and behaviour of individuals must be sacrificed and curtailed for the good of the whole.

This idea of individual restraint and sacrifice in the face of a larger organic whole is also expressed by Callicott when he says:

> Our organic health and well-being, for example, requires vigorous exercise and metabolic stimulation which cause stress and often pain to various parts of the body and more rapid turnover in the life-cycle of our individual cells. For the sake of the person taken as a whole, some parts may be, as it were, unfairly sacrificed. On the level of social organisation, the interests of society may not always coincide with the sum of the interests of its parts. Discipline, sacrifice, and individual restraint are often necessary in the social sphere to maintain social integrity as within the bodily organism.[35]

Charles Birch, another holist environmental writer, campaigns for such social integrity in this manner:

> The life force...is calling humanity to a new organisation of human societies. Here is where the integration is most urgently needed. Here is where integration is most incomplete and inadequate.[36]

Birch goes on to identify exactly how we may characterise any form of behavior or action that does not serve integration: they are instances of evil:

> Evil is always the assertion of some self-interest without regard to the whole, whether the whole is conceived as the immediate community, the total community, or the total order of the universe.[37]

These flirtations with the supremacy of the organic integrated whole are, indeed, a problem for other investigators of environmental and ecological theory. Most notably, the idea of organic unions in nature and society has given rise to the charge of social fascism. Janet Biehl, in particular, makes strong connections between ecological holism and the social fascism exemplified by Nazism.[38] Whether or not Gaian philosophy can and will feed into the support for political and social fascism is yet to be determined but Biehl points out (with regard to the organic eco-spirituality ideas that have permeated the modern German ultra-right) that a real danger may be evident. She indicates that within this ultra-right Rudolph Bahro's return to the Volkisch Spiritualism is actually a philosophical alignment to Nazism. Biehl goes on to quote Bahro's calls for a 'Green Adolph' when Bahro talks of the need for a new fuhrer to drive the German people towards environmental friendliness. Rudolph Bahro was an acknowledged leader of the German Green movement and his writings have, as Biehl points out, became riddled with statements like the following:

> the most important thing is that...[people] take the path 'back' and align themselves with the Great Equilibrium, in the harmony between human order and the Tao of life.[39]

This sort of sentence also pervades Goldsmith and Capra but to label their work as neo-Nazi might be to exaggerate the point.

Michael Zimmerman echoes Biehl's concerns over social unity and harmony ideals when he says:

> The fact that National Socialism was a perverted expression of the desire for social harmony does not make such desire illegitimate, but does require that critics of Modernity--including counter-culturalists and radical ecologists--proceed carefully in calling for alternatives.[40]

Biehl's concern about the re-emergence of holist ecological ideas and neo Nazism might be a little extreme but if we adjust them to say that if holist environmental concepts are manifested strongly in social situations--as for instance in Goldsmith's social systems--then elements of fascism do become apparent.

Blaming any particular facet of Nazi policy and thought as a derivative of nothing else but organic views of nature and society is a far too clumsy and simplified way to undertake scholarship, of course. Many social thinkers would doubt that philosophies of nature have anywhere near a sole determining affect on the character of social operations.[41] Having said this, it wouldn't be wrong to believe that philosophies of nature might serve as an aid in the legitimisation of political ideas (by either the political elite, the politically-aspiring, or their fanatical audiences) at certain times. Whether this legitimisation is absolutely necessary, or of partial importance, or of no relevance at all, to the social situations that the philosophy is supposedly connected with, is, of course, a subject unresolved. However, we can at least note here that there are some situations in which philosophical worldviews do at least offer profound legitimising schemes for particular political viewpoints and actions and that it is wholly feasible to regard it as possible that a unitarian philosophy of nature can act as a stimulus to the theoretical formulation of social systems that might possess fascist tendencies.

However, the point that I want to press here is that my position in all of this is slightly different from those like Biehl who claim that unitarian thought can lead to fascist political and social situations. My view is that we do not have to wait for a Nazi-like social situation for unitarianism to rear its head in ultra-conservative ways. The concept of a domineering system is not something only inherent in Nazism and other such fascist social unities, it is a concept that is alive and well in the liberal democratic society of the modern world. As such, it might be noted that the holistic thought which is said by Biehl to have rationalised Nazism, is now rationalising liberal capitalism. What this means is that although theoretical caution may salvage environmental holism from moving towards policies of a Nazi-esque nature, the greatest ideological danger of unitarianism may be its operation through the more politically acceptable medium of the 'system' concept. This important point is where the next chapter starts off.

Natural Conservatism: Summary and Conclusions

According to many modern day philosophers of nature, humanity must learn to acknowledge unity in the natural world (what Capra, calls 'The Web of Life' and Lovelock calls 'Gaia') and also behave as though this knowledge was important (and act according to what Goldsmith calls 'The Way'). One of the ways of acknowledging this unity--and of acting consistently with The Way--is, suggests Goldsmith, to accept the continuity between nature and humanity, between natural and social systems.

If Goldsmith had his way, humans would all mega-mutate towards The Way; towards Gaian ethics. Society would become traditional, less modern, and in doing so would behave within a Gaian order, thereby becoming less destructive towards nature. In the process they would also become entrenched with the metaphysics and values of unity, hierarchy and functionalism so that, in a society like the one Goldsmith has planned, individuality, dissension and difference may very well be devalued.

Such neglect of individual differences and dissension is evident, according to many social commentators (including Bauman, Lyotard and Layder) in the social theory of those from whom Goldsmith draws his theories, i.e.: structural functionalists like Talcott Parsons. Both Goldsmith and Parsons draw a lot of their inspiration from holist biological ideas, and just as these ideas are anti-individual and intolerant of difference and change in the biological world (as has been shown in Chapter Two), so they are anti-individual and intolerant of difference and change in the social world.

One final point might be made here. Perhaps Goldsmith is right in that his traditional societies might be less ecologically destructive than modern societies but this has nothing to do with Gaian holism, social hierarchy and social functionalism as Goldsmith would argue. It is due to the less destructive material potential of such societies. Such societies, because they are small-scale and needs based rather than large-scale and wants based, would be less likely to glorify and celebrate technological and economic gain and this in itself would confer upon them a greater degree of eco-friendliness than modern day society. However, a non-stratified, non-hierarchical, non-holistic, anarchic society of the scale Goldsmith posits for his planned ecological societies would be just as light on the environment as those replete with the hierarchy and holism of Gaian ethics.

Notes for Chapter 4 (Natural Conservatism)

1. Note, for instance, the following comments of Goldsmith's:

> "Ecology reflects the values of the biosphere." (from Goldsmith, *The Way: An Ecological Worldview*, NY: Shambhala, 1993, p.82)
>
> "There is no fundamental barrier separating man and other living things." (from Goldsmith, *The Way*, Ch.37.)
>
> "The biosphere displays a...single set of laws whose generalities apply equally well to biological organisms, communities, societies and ecosystems and to Gaia herself" (from Goldsmith, *The Way*, p.186).
>
> "There is no fundamental difference between the structure and behaviour of vernacular [i.e. traditional] man and that of other living things. Both are governed by the same laws that govern the natural systems which make up the Gaian hierarchy" (from Goldsmith, *The Way*, p. 209).

2. J.B. Callicott "Animal Liberation: A Triangular Affair", in R. Elliot, ed, *Environmental Ethics* (Oxford: Oxford University Press, 1995) p.41.
3. R.U. Ayres, *Information, Entropy and Progress* (Woodbury, NY: AIP Press, 1994) p.186.
4. See: F.B. Golley, *A History of the Ecosystem Concept in Ecology* (New Haven: Yale University Press, 1993); E.P. Odum, *Fundamentals of Ecology* 3rd edn (Philadelphia: Saunders, 1971).
5. R.U. Ayres, *Information, Entropy and Progress*, pp.186-187.
6. L. von Bertalanffy, "General System Theory", in N.J. Demerath & R.A. Peterson, eds, *System, Change and Conflict* (NY: Free Press, 1967) p.117.
7. Bertalanffy, "General System Theory", p.118.
8. J. Hagen, *An Entangled Bank: The Origins of the EcosystemConcept* (New Brunswick: Rutgers University Press, 1992) p.69.
9. Bertalanffy, "General System Theory", p.118.
10. Ayres, *Information, Entropy and Progress*, p.209.
11. *ibid*
12. As has been outlined in Chapter Two with reference to D. Miller, "Metaphor, Thinking and Thought", *ETC.: A Review of General Semantics*, 39, 134-50 (1982); and as will be outlined in Chapters Five and Six with reference to F. Nietzsche, "Philosophy and Truth" in D. Breazeal, ed, *Selections from Nietzsche's Notebooks of the Early 1870s*, (NJ: Humanities Press, 1979); G. Lakoff and M. Johnson, *Metaphors We Live By* (Chicago: University of Chicago Press, 1980).
13. See P. Mirowski, ed, *Natural Images in Economic Thought: Markets Read in Tooth and Claw* (Cambridge: Cambridge University Press, 1994)

14. I am by no means the first to suggest this. See, for instance, Mirowski, *Natural Images in Economic Thought* and R. Young, *Darwin's Metaphor* (Cambridge: Cambridge University Press, 1985).
15. R. Young, (1992) "Science, Ideology and Donna Harraway", *Science as Culture*, 3, no. 2, (1992) p.167.
16. T. Benton, *Natural Relations: Ecology, Animal Rights and Social Justice*, London: Verso, 1993, p15.
17. We can illustrate this coarsely by noting the influences upon one of today's foremost unitarians; Edward Goldsmith. Goldsmith says that both human communities and ecological communities function like systems. Goldsmith himself claims to have been influenced by the sociologist Talcott Parsons, the physical scientist James Lovelock, and the ecologist, Eugene Odum, when coming to this conclusion. But Parsons was influenced by social thinkers that themselves were holist-inspired ex-biologists, people like Walter Cannon and Lawrence Henderson. Parsons was also influenced by people who themselves were directly influenced by biological ideas of holism, people like Emile Durkheim, Herbert Spencer and Vilfredo Pareto (as recounted in: R. Lilienfeld, *The Rise of Systems Theory: An Ideological Analysis*, Krieger, 1988; S.J. Heims, *The Cybernetics Group*, Cambridge: MIT Press, 1991; G.P. Richardson, *Feedback Thought in Social Science and Systems Theory*, Philadelphia: University of Pennsylvannia Press, 1992; S. Heyl, "The Harvard Pareto Circle", (Reproduced on the *Science as Culture* website). Goldsmith's other influences, Lovelock and Odum, on the other hand, were influenced by people like the biologist, Bertalanffy, and the systems engineers, John von Neumann and Norbert Wiener (as recounted in J. Hagen, *An Entangled Bank: The Origins of the Ecosystem Concept*, New Brunswick: Rutgers University Press, 1992; E. Goldsmith, *The Way: An Ecological Worldview*, NY: Shambhala, 1993; F.B. Golley, *A History of the Ecosystem Concept in Ecology*, New Haven: Yale University Press, 1993, F. Capra, *The Web of Life*, (London: HarperCollins, 1996). But Bertalanffy, von Neumann and Wiener themselves were influenced by people like Cannon, Henderson, Parsons and other social holist theorists (as recounted in N. Wiener, *Cybernetics, or Control and Communication in the Animal and Machine,* Cambridge: MIT Press, 1961); M. Davidson, *Uncommon Sense: The Life and Thought of Ludwig von Bertalanffy*, LA: Tarcher, 1983; and F. Capra, *The Web of Life*.) This group of people were, of course, influenced by holist biologists. Out of this inter-articulation of influences comes a morass of dead and dying metaphors which Edward Goldsmith (and others) have forgotten are metaphors.
18. See Lilienfeld, *The Rise of Systems Theory*; Heims, *The Cybernetics Group*; Heyl, "The Harvard Pareto Circle"
19. Lilienfeld: *The Rise of Systems Theory*.
20. A. Clayton, "Systems Theory: Some Caveats", *Environmental Values*, 2, (1993) p. 159.
21. Darhendorf as quoted in J. Farganis, *Readings in Social Theory: The Classic Tradition to Postmodernism* (NY: McGraw-Hill, 1993) p.222.
22. D. Layder, *Understanding Social Theory* (London: Sage, 1994), p23.
23. Layder, *Understanding Social Theory*, p.14.

24. P. Dickens, *Society and Nature: Towards a Green Social Theory*, (London Harvester Wheatsheaf, 1992) p.44
25. Goldsmith, *The Way*, p.112.
26. Dickens, *Society and Nature*.
27. J-F. Lyotard, *The Postmodern Condition*, transl: by G. Bennington and B. Massumi, (Manchester: Manchester University Press, 1984) p.xxiv.
28. Z. Bauman, *Intimations of Postmodernity*, (London: Routledge, 1992) p.39. As the following critics of structural functionalism, have noted, in such a system as Bauman describes, even change (and the desire for change) is co-opted to give rise to stability:

    "In countries with liberal or advanced management the struggles and their instruments have been transformed into regulators of the system" (from Lyotard, *The Postmodern Condition*, p.13).

    "Within the structural functionalist school...[h]ere conflict is often seen as leading to ultimate harmony and stability." (from N.J. Demerath, "Synechdoche and Structural Functionalism", in N.J. Demerath and R.A. Peterson, eds, *System, Change and Conflict*, NY: Free Press, 1967, p501).

    "Deviant behaviour is regarded as a temporary aberration, a failure of the socialisation process rather than an expression of difference or dissent" (from Farganis, *Readings in Social Theory*, p.364).

    If even conflict against the system can be functionalised within the system to contribute to stability, then one may wonder what social phenomena cannot possibly be incorporated into the social system. This is the main problem with the Parsonian social system; at least according to it one of its prime attackers, Ralf Dahrendorf. For example, Farganis paraphrases Dahrendorf as saying that within a Parsonian social system:

    "Everything is too neatly laid out; the family performs the reproductive function and replenishes society with fresh births; the educational system secures conformity and adherence to the rules of thought, its function as an agent of socialisation, and the division of labour allocates different roles that people must play in a complex economic system." Farganis, *Readings in Social Theory*, p.222.
29. Goldsmith, *The Way*, p.216.
30. C.D. Rollo, *Phenotypes: Their Epigenetics, Ecology and Evolution*, (London: Chapman and Hall) p.28.
31. L. Westra, *An Environmental Proposal for Ethics* (Baltimore: Rowman and Littlefield, 1994), p.126.
32. *ibid*, p127.
33. Goldsmith, *The Way*, p.369.
34. *ibid*, p.8.

35. Callicott, "Animal Liberation: A Triangular Affair", p.42. Michael Zimmerman also notes many similar remarks within Callicott's other writings. See M.E. Zimmerman, "The Threat of Ecofascism", *Social Theory and Practice*, 21 (1995) :207-238.
36. C. Birch, *On Purpose*, (Sydney: UNSW Press, 1990) p.108.
37. *ibid*, p15.
38. See J. Biehl, "Ecology and the Modernization of Fascism in the German Ultra-Right", *Society and Nature*, 2, (1994): 21-78; J. Biehl, and P. Staudenmaier, *Ecofascism: Lessons from the German Experience*, Edinburgh: AK Press, 1995). See also Zimmerman, "The Threat of Ecofascism."
39. Biehl, "Ecology and the Modernization of Fascism in the German Ultra-Right", pp.152
40. M.E. Zimmerman, *Contesting Earth's Future: Radical Ecology and PostModernity* (Berkeley: University of California Press) p.59.
41. Examples of scholars that have dealt with the interactions of worldviews and social practices but have also maintained that these interactions are never played out so that a natural philosophy maps exactly on to a social practice include: D. Bloor, *Knowledge and Social Imagery* (London: Routledge and Kegan Paul, 1976), J.C. Greene, *Science, Ideology and Worldview: Essays in the History of Evolution Ideas* (Berkeley: University of California Press, 1981), R. Young, *Darwin's Metaphor*; B. Barnes and S. Shapin, eds, *Natural Order* (London: Sage, 1979); W. Kempton *et al, Environmental Values in American Culture* (MA: MIT Press, 1995).

CHAPTER 5

Uniting the Ecosystem with the Economy

Introduction: The Rise of Cybernetics

In the middle of the Twentieth century the unity of nature was being reinterpreted using the concept of the 'system'. Nature, viewed as a system, was thought to be a vast interlocking web of energies, entities, actions and purposes, and to this web of interacting components, humanity, in some way or another, belonged.

This notion of the interacting, interlocking system is at the forefront of modern day manifestations of the unity concept, so much so that expressions about the unity of nature nowadays can very rarely escape the ideas and terminological practice of systems science. Even when examining the philosophy of a late Twentieth century holist environmentalist like Arne Naess--who, unlike most of the intellectuals identified in this work, has no real interest in systems--we still see within him the influence of systems science. For instance, when explaining Deep Ecology's conviction toward ecological unity, Naess characterises ecological unity as: "a multiplicity of more or less lawful, interacting factors may operate together to form a unity, a system."[1] This cornering of the unity of nature market by systems theory is not total but it is hard to understate. For this reason systems science is often regarded as the science of unity.

One of the obvious ways to note the relationship of contemporary systems science to conceptions of natural unity involves the very words 'system' and 'unity'. The closeness of the word 'system' to the conceptual character of 'unity' can be explored by perusing the definitions of those that first scientised the systems concepts. Ludwig von Bertalanffy, for instance, defined systems as "complexes of elements standing in interaction."[2] Thus when we are talking about those that first formulated scientific theories about systems, and Bertalanffy is adjudged by most systems theorists of today to have had a prominent role, then the idea of unity was a full-time metaphysical commitment when systems theories were first outlined.

Given this history of the interaction between unity and systems (whose details will be entered into in the sections that follow) it is of little surprise to find that--just like unitarians who cannot divorce themselves from systems thinking–systems thinkers can not divorce themselves from unity notions. For instance Schulze and Zwolfer define a system thus:

> a system may be defined as an integrated entity whose overall properties are different from the properties of its elements.[3]

This definition lies very close to the holistic definitions of unity used by unitarians (like Naess above) who do not directly employ systems theory.

These definitions of what it takes to be a system are, of course, rather open and vague; so much so that the system concept might be thought of as encapsulating just about any object, any process, any thing that exists in space and time. The description of a thing as a system thus seems rather arbitrary.

This willingness to vaguely define systems--to draw upon the open-endedness of the system concept--is actually a great source of strength for systems advocates. They may happily apply their concept, and its attendant scientific and philosophical rules and principles, to all manner of things and processes without much disagreement coming from within science that they cannot do this because there is no scientific, grounded definition of what a system can and cannot be. Any previously imagined unity (such as a society, forest, nation, household) can now become a system and be submitted to a systems analysis.

Here some readers might spot some sort of paradox between systems theory's veneer and its substance. Systems science has gained a lot of its adherents, support, funds, and popularity because it was perceived to be ultra-modern and ultra-scientific, yet it is actually not very scientific at all, merely a play towards certain philosophical and metaphysical biases. Although it is a discipline that uses and produces a lot of numbers, a lot of mathematics, a lot of diagrams and a lot of sophisticated terms, systems science nevertheless is a science indissolubly welded to peculiar metaphysical standpoints. It merely hides this metaphysics under mathematical equations, flow diagrams and jargon.[4]

Whether systems ideas are actual scientific concepts or merely the philosophical musings of scientists, the major use of the system concept to unity fans is its glossy scientificity. The scientificity of systems theory has a lot to do with its supposed discovery and elaboration of a universal principle. This is to say systems theory claims to outline the very process that shows exactly how nature is united.

This mechanism--this glue that unites everything–is called 'negative feedback'. Systems are systems, according to the systems theorists, because of the operation of negative feedback between the systems various components.

Feedback may be defined as the circular process in which the activity of a thing contributes to the activity of another thing which then feeds back to contribute the activity of the first thing. When the activity of one thing influences the activity of another thing in a negative way, i.e. when an activity is dampened down or restrained, we get what systems theorists regard as the magical process of negative feedback. A simple example of negative feedback would be the thermal feedback system in a refrigeration unit where the temperature of the internal compartments are monitored by thermal sensors which automatically switch the refrigerator motor on when the sensors detect that the compartment has risen above a certain temperature. When the compartment then becomes cold enough, the sensor detects this and switches the motor off. All this keeps the compartment at a nice stable cool temperature and saves us humans from tediously switching on and off the refrigeration unit many times a day. Because all the components are integrated together and are functioning in an interdependent way, the refrigeration unit as a whole can be regarded as a holistic united entity. Take away this interdependent functioning, either in practice by damaging the motor, or in theory by studying the refrigerator after reducing it to its parts, and the entity will break down and/or cease to be understood properly.[5]

According to its advocates, the global Gaian system is a superb living example of the operation of negative feedback. For Lovelock and the other Gaians, negative feedback operates between the geobiological components of the Earth to keep the atmosphere and the seas in thermal and chemical constancy. In such a system, there are many things, many activities, and many sensors but, according to Lovelock, the principle is just the same.

Lovelock, however, was not the first who aspired to bring system concepts into the ecological realm. Drawing his insights from Norbert Wiener's mathematics of feedback, Evelyn Hutchinson[6] was one of the first ecologists to emphasise the role of negative feedback in ecological situations. Hutchinson's unit of feedback was not the whole earth, as in Lovelock's case, but the ecosystem. Hutchinson used the process of feedback to account for the stability of ecosystems:

> It is well known from mathematical theory that circular paths often exist which tend to be self-correcting within certain limits, but which breakdown, producing violent oscillations, when some variable in the system...transgresses limiting values. It is therefore, usual to find in nature's systems various mechanisms acting to damp oscillations.[7]

Such announcements of the reality of ecologically-related feedback processes continue in modern day ecosystem ecology and its inescapable importance as an ecological concept is exhorted by Edward Goldsmith too:

> the operation of all sorts of internally-generated negative feedback mechanisms (Eugene Odum's environmental hormones) which inhibit the growth of species that are displaced in the succession towards a climax is clearly visible to all but the most prejudiced eye.[8]

It might be alleged that systems theory is far more than just finding worldly examples of negative feedback (and that systems ecology is more than just finding ecological examples of negative feedback) but although these allegations may be sound enough it is also true to say that both systems theory in general and systems ecology specifically can hardly exist without the concept of feedback.

If we trust a systems theorist to elaborate upon the history of their subject they would surely relate it to the rise of the discipline known as 'cybernetics'. Cybernetics, they would point out, first self-organised itself into an intellectual system of thought within a military background during 1940s America. Hagen, who is not an avowed systems theorist, expresses this history in a concise way thus: "cybernetics had developed largely out of the wartime problem of designing an automatic control system for anti-aircraft guns."[9]

The two leading figures of cybernetics, John von Neumann and Norbert Wiener, were themselves both involved in the military development of cybernetic machines during and immediately after the Second World War, although:

> whereas von Neumann remained a military consultant throughout his career; specialising in the application of computers to weapons systems, Wiener ended his military work shortly after the first Macy meeting.[10]

The Macy meetings were a short series of conferences convened by early systems theorists to discuss the emerging problems and ideas of their field. About these conferences, which both von Neumann and Wiener attended, Fritjof Capra writes:

> The conceptual framework of cybernetics was developed in a series of legendary meetings in New York city, known as the Macy Conferences. These meetings--especially the first one in 1946--were extremely stimulating.[11]

Capra has enormous respect for von Neumann and Wiener and strongly advocates them as profoundly influencing the development of unitarian ideas in both biological and ecological thought. Certainly, within the science of ecology, cybernetics is richly represented in ecosystems studies and in systems ecology. Although Wiener said nothing about ecology, many ecological ideas within systems ecology, including James Lovelock's Gaia, were developed as variants inspired by the cybernetics of Wiener.

Von Neumann himself also said little or nothing about ecology directly but he was, says Capra, given over to philosophising about general systems in "biology and appreciated the richness of natural, living systems."[12] Von Neumann's work, too, is said by Capra to have been a direct influence upon later systems ecologists like, for instance, Eugene Odum, Howard Odum and James Lovelock. Indeed it might be pointed out that the Odums and Lovelock owe more to cybernetics than to any traditional approach in ecology. This is evident (and, incidentally, approvingly acknowledged by Edward Goldsmith) when Odum declares:

> ecosystems can be considered cybernetic in nature, but control functions are internal and diffuse rather than external and specified as in human engineered cybernetic devices.[13]

Similarly Lovelock states "Gaia has been...the hard science view of a physical chemist with an interest in control theory"[14]; 'control theory' being another name for those principles that describe feedback tendencies in cybernetic systems. Admissions like this permeate Lovelock's work so that we must judge that Gaia is not merely imbued with feedback/control theory but is a product of the intellectual heritage that is cybernetics and systems science.

When looking at the systems ecology of Eugene and Howard Odum it's noticeable that they spent their whole working lives analyzing and articulating feedback processes for the chemical transfer patterns in the physical environment of ecosystems. The Odums protracted elaboration of carbon cycles, nitrogen cycles and phosphate cycles are examples of such chemical negative feedback and transfer patterns. In Lovelock's Gaia theory the feedback processes that he has observed included the cycling of these materials but the cycles are postulated on a global scale. Capra, too, makes much of the ecological and philosophical importance of the negative feedback cycles involving carbon, nitrogen, phosphate etc, in his popular books. He then goes on to compare the prevalence of negative feedback to positive feedback in ecology:

> Indeed, purely self-reinforcing phenomena are rare in nature, as they are usually balanced by negative feedback loops constraining their runaway tendencies...In an ecosystem, for example, every species has the potential of undergoing an exponential population growth, but these tendencies are kept in check by various balancing interactions within the system.[15]

This example, incidentally, is one of the favourites for those with an interest in both ecology and systems theory for not only does it give an easily understandable example of the presence of negative feedback it also does other important things for cybernetics. Firstly, it links cybernetics not just with any old science but with the politically attuned science of ecology; thus promoting cybernetics as a politically agreeable and socially-important science. Having positioned itself within that science, systems theory can more easily lose its somewhat politically-disagreeable connection with weapons of war. Systems theory--via systems ecology--then goes on to make very acceptable claims about the reality of ecological situations (i.e. nature is in unity, nature is in balance, etc). These claims are acceptable because they appear to restate what has been known for so long.

Systems Ecology as Management Ecology

If systems ecology can be characterised as 'cybernetics meets ecology' then the two men that encouraged that meeting most, and derived most from it, were the Odum brothers. As introduced previously, the Odum brothers comprised Eugene, who obtained his Ph.D. not in ecology, but physiology, and Howard, who obtained his postgraduate training in meteorology. The Odums were able to gain support for systems ecology for two reasons. Firstly, the United States Atomic Energy Commission (the AEC) wanted to know what affects radiation had on human health and the health of human environments. Eugene Odum was the first scientist to impress upon the AEC that systems studies of ecological settings were the best way to do this and he prospered because of it.

Before the patronage of the AEC towards the work of Eugene Odum, ecology (whether it was community, ecosystem or population ecology) was generally only funded by university grants. As Chunglin Kwa points out: "in this context the importance of the AEC as a patron of 'big ecology' must be understood".[16]

The second reason for the support of Odumian systems ecology in the 1950s and 60s was its perceived relevance to environmental and ecological management. Kwa explains that systems ecological concepts had "fallen on fertile ground because of their management applicability"[17]. As Kwa explains it; systems ecology was not only new and quantitative (and produced numbers that management could use in planning) it also boasted the language and the rationality that managers (and their overseers) liked. Not only that but its results were presented in forms very much the same as management studies were presented. Ecosystem flow charts resembled the flow charts of 1950s management techniques so much that one could just plug the ecological results into resource management charts without fear of rejection. In this way, systems ecology became management ecology. Biggins found these links so remarkable that he was compelled to comment that the central metaphors of modern ecological theory (i.e.: control, regulation, interaction, feedback, flows, etc) derive from post-Keynesian capitalist management techniques and computing.[18]

This idea that systems ecology was management ecology was not discouraged but positively fostered by the Odums and other ecosystem scientists. Eugene Odum, for instance, is said by Frank Golley as having believed that ecosystem science could provide a "pure science basis for landscape planning in the future."[19]

As with the management of human systems via systems management, the management of natural systems via systems ecology was, as Golley would put it, 'fundamentally economical'. Nature, like society, could be converted into an economic entity. It was hardly any trouble at all to reconfigure the carbon cycle or the nitrogen cycle into models of the economy. Indeed this is what the Odums have continuously sought to do.[20]

Kwa points out that:

> The development of systems ecology was a manifestation of great technological optimism with regard to the possibility of the management of natural systems.[21]

This optimism, suggests Kwa, was developing in the 1950s and 1960s at the same time that there was a general perception of the applicability of systems analysis to every form of human endeavour. However, such optimism for ecological systems management can hardly be periodised as being just a phenomena of the 50s and 60s. Nowadays the 'systems ecology equals management ecology' idea is alive and well in the form of 'environmental integration management'.

Within this brand of environmental management it is often thought that only an integrated systems approach can solve problems in the environment (yet it merely reduces these problems to something that looks like it can be solved). Just as systems ecologists often castigated small-scale approaches as being fragmentary approaches to environmental management in the 1960s, so environmental managers nowadays castigate less than total perspectives in today's management of the environment. Calling for a total ecological appraisal, or worse claiming that such an appraisal exists and then using it as such, is still part of the tradition that mistakes systems in the natural world for the whole of the natural world. This is not to say systems approaches are in no way useful when they focus upon the subjects of interest to them (i.e. energy flows, material pollution, productive capability, etc) but it does suggest that the items of interest to systems managers are not always the items that should be of interest in environmental protection (such as the well-being of individual species and the communities they make up).

Although the success of ecosystems analysis was debated within the ecology profession itself during the 1950s, 1960s and 1970s, many non-ecological scientists, including biologists, did not doubt its efficacy and applicability and also its scientific hardness. This is still the case in ecological pedagogy where systems ecology receives extended coverage in general science syllabi. Ecosystem studies have thus become a potent, if not dominant, way of looking at nature; this dominance being at least partly encouraged by its links to managerial issues.

Another interesting twist of the 'systems ecology equals management ecology' equation is also worth mentioning. According to Golley, George van Dyne, the systems ecologist that led a section of the grandiose 1968-72 International Biological Program, ran his Grasslands Ecosystem project within this Program as a 'typical systems manager'. Like a "dogmatic" king overlooking his hierarchically-organised domain, van Dyne "fiercely...drove his colleagues, co-workers and graduate students hard" towards maximum project efficiency.[22] If nothing else this tells us about the similarity of ecosystems management to the management ideas of the time but it may also hint at more. It might expose the ease to which a metaphysical viewpoint may totally dominate the thoughts of its practitioners so that nothing in the world can be conceived or practiced without regard to it.

Order out of Chaos: the 'New Sciences'

For many systems theorists in this turn of the millennium period, systems theory is being superseded by something new. This something new is 'complex systems theory'. No more do systems theorists deal with just simplified things. They have matured in their way of thinking to believe that everything is complex, even the simple things.

The difference between systems theory and complex systems theory, according to Felix Geyer[23], is that where systems theories were implicitly based on a mechanistic or Newtonian clockwork image of the universe, which had a preference for order and stability, complex systems theory appreciates the intimate connection between order and disorder and between stability and instability.

This might make an outsider believe that systems scientists have admitted, at last, to being unable to model and explain all things in nature using only one methodological framework but that is not necessarily the case. In fact, armed with complex systems theory, some scientists now believe that they may be able to explain and model just about anything in nature.

Both systems theory and its descendent, complex systems theory, evoked over-excitement when they were respectively emerging on the international academic scene. Just as systems theorists during the post-War period spoke of their discipline in revolutionary terms[24], so complex systems theorists speak of complex systems theory in revolutionary terms some forty years later.[25]

It should be noted that complex systems theory goes by various names. It is, perhaps, more familiarly known by its constituent parts; complexity theory and Chaos theory. Added to these two core theoretical nodes might be number of others, like the self-organisation theories of Maturana, Varela, Prigogine and Jantsch[26] and also Lovelock's Gaia theory. All these four bodies of theory are all in some way related, having in common, according to Capra, key concepts such as: "chaotic attractors, fractals, dissipative structures, self-organisation, and autopoietic networks."[27] When spoken about together, complexity theory, Chaos theory, self-organisation theory and Gaia theory, are often dubbed the 'New Sciences'. To save time, this is what I shall call them too.

To sum them up we might say that the theories contained within the New Sciences advocate an evolutionary, dynamical systems view of the cosmos that elevates processes over substances and recognises the ever-present action of self-organising order emerging from chaos. Or as Davies puts it, the New Sciences tell:

> the story of the universe as one of increasing complexity and organisation emerging spontaneously from primordial simplicity and uniformity. The self-organising and self-complexifying power of the laws of physics which are only now being studied, constitute a second remarkable property.[28]

True to the intellectual marketing zeal of those who promote complex systems theory, both Chaos theory and complexity theory have independently been hailed as revolutionary sciences. Capra, for instance, says of complexity theory:

> The discovery of this new mathematics of complexity is increasingly being recognised as one of the most important events in 20th century science.[29]

And likewise, of Chaos theory, Hayles declared:

> it is already apparent that Chaos theory is part of a paradigm shift of remarkable scope and significance.[30]

The word 'chaos' would seems to conjure up things anarchic, things sporadic, things random, yet the chaos in Chaos theory is not really like this. As various advocates of Chaos theory put it:

> Although chaotic behaviour is, by definition, dauntingly difficult to model, there is still some underlying order in its manifestation.

In Chaos theory the term chaos has acquired a new technical meaning. The behaviour of chaotic systems is not merely random but shows a deeper level of patterned order.[31]

The chaos in Chaos theory is usually ascribed a creative role. New Scientists cast it as an agent of ongoing ordering in the natural world:

Disorder can play a critical role in giving birth to new, higher forms of order.

Out of chaos comes forth the fertile variety of forms of existence and life in the universe. Chaos is the father of innovation.[32]

The order that is hidden in, and emerges from, chaos is not of a simple kind but is, according to Chaoticians, extraordinarily complex. Hence the name of the theory that explores this type of order is 'complexity theory'.

The process whereby complex order emerges from chaos is often given the title 'self-organisation', the maintenance of which is termed 'self-regulation'; two terms that we have come across occasionally in previous chapters. Within the New Sciences, the study of such ordering and regulation becomes so intense it is often described as constituting a particular theoretical node: 'self-organisation theory'.

Chaos theory and self-organisation theory are intimately linked (so say the New Scientists), since "chaos is the basis of the ability of living matter to self-organise itself".[33] Self-organisation, too, is also presented in paradigm-changing terms by certain New Scientists:

I have taken some time to outline the emerging theory of self-organising systems because it is today the broadest scientific formulation of the ecological paradigm with the most wide-ranging implications.[34]

Life, itself, we are told by Capra, is a system defined by its spontaneous self-organising properties. By living systems Capra means not only individual organisms but living systems at higher levels:

like individual organisms, ecosystems are self-organising and self-regulating systems.[35]

Chaos theory, complexity theory and self-organisation theory have also been intimately linked to the unity of nature. Complexity theory, for instance, has often been intellectually tied with the unity of nature in general and with the Gaia theory in particular. Witness this statement from Lewin as an example:

Today, researchers are viewing Gaia and the whole notion of the superorganism in light of the modern mathematical theory of complexity.[36]

Paul Davies, too, finds that:

Gaia provides a nice illustration of how a highly complex feedback system can display stable modes of activity in the face of drastic external perturbations. We see again how individual components and sub-processes are guided by the system as a whole to conform to a coherent pattern of behaviour...The fact that life acted in such a way as to maintain the conditions for its own survival and progress is a beautiful example of self-regulation. It has a pleasing teleological quality to it.[37]

Similarly, both Fritjof Capra and Fred Pearce, the latter a regular writer for *New Scientist*, the popular science magazine, are also impressed by the relationship of Gaia theory to self-organisation and complexity:

James Lovelock had an illuminating insight that led him to formulate a model that is perhaps the most surprising and beautiful expression of self-organisation--the idea that the planet Earth as a whole is a living, self-organising system.

The ultimate superorganism is Gaia. This links up to the new world of chaos and complexity theory. This theory holds that within complex systems, order can spontaneously emerge out of chaos. That fits exactly with what Lovelock sees Gaia as being.[38]

As well as Lewin, Davies, Capra and Pearce, two more prominent spokespeople for Chaos theory, Cohen and Stewart, also admit to appreciating Gaia theory by saying:

Gaia, as an integrated dynamical system, replete with feedback loops and stabilising subsystems, is an entirely respectable concept.[39]

Whether or not it is actively linked to the Gaia theory, the complex systems view in the New Sciences still advocates a united conception of life and the universe:

Chaos in a certain sense reveals the unity of the universe and the hidden tie between one thing and another.[40]

Merry goes on to write that a:

> basic characteristic of complex systems is connectivity. In a world of complex systems everything is connected to everything else.[41]

Similarly, Fraser Clarke says of Chaos theory that it is:

> unification science, interrelation science, whole science. It shows you that there is a connection between everything.[42]

The 'New Sciences' and Ecology

Fritjof Capra indicates that one of the reasons he finds:

> the theory of self organising systems so important is that it seems to provide the ideal scientific framework for an ecologically oriented worldview.[43]

Readers of Capra's works who consider themselves to be active environmentalists, as opposed to New Age spiritual ecologists, might like to ask him about the actual connections between the New Sciences and environmentalism, since details of the connection are not self-evident in his books. Although Capra talks about the New Sciences as prompting a paradigm change to an 'ecological' and 'environmentally-friendly' worldview, whereabouts does he actually talk of the world's environmental problems in any detail? It seems, nowhere. As an example consider Capra's book, *The Web of Life*, in which Capra tries hard to attach himself not only to ecological science in general, via references to ecosystem ecology, but to Deep Ecology specifically, a popular environmental movement:

> The sense in which I use the term 'ecological' is associated with a specific philosophical school, and, moreover, with a global grassroots movement, known as 'Deep Ecology' which is rapidly gaining in prominence.[44]

Now Deep Ecologists, themselves, might be puzzled at Capra's attempt to align his 'ecological' approach in *The Web of Life* with Deep Ecology philosophy, since it is a book about the philosophy of complex systems theory and not one which seeks to explore environmental problems in relation to any of the precepts of Deep Ecology.[45] Capra's retort to this would probably go something like this: 'the cause of the environmental crisis is our modern worldview and what we need in order to save the world is a new worldview which posits the unity of all things'. Herein lies the connection to Deep Ecology but it might strike many Deep Ecologists as being a very shallow and superficial one since all it seems to suggest is that both Capra and the Deep Ecologists share an intellectual commitment towards a change in worldview and a metaphysical commitment to unity.

Although he declares at the beginning of his book that he is a Deep Ecologist, nowhere else in his book does he take up any of the debates raised by Deep Ecology nor does he enter into the value premises and focal arguments (such as biospheric egalitarianism, anti-classism, population and pollution control) of the Deep Ecologists.[46] In fact, Capra does not even go anywhere near the literature of this 'world-wide grassroots environmental movement' he professes to be a part of. Given this, the overtures he makes to Deep Ecology must be interpreted as a play for political legitimacy within the environmental movement. His ideas, so Capra thinks, must be given cognizance by environmentalists, if for no other reason, than he is expressing their philosophy of nature. Many other New Science advocates suffer from a similar lack of credibility with regards to their capacity to engage with actual environmental problems in their writings. However, as they are regarded as the philosophical or metaphysical wing of the environmental movement, they, like Capra, are generally forgiven.

Another point to make about the New Sciences and their connections to ecology revolves around the possible attachment of ecology to Chaos theory. Dealing with chaotic phenomena as it proclaims, Chaos theory, if applied directly to ecology, might be thought to be somewhat more compatible to the disordered forest ecology theorised by Gleason than the superorganicism of Clements. However, this is not the case. Chaos theory, as Capra, Davies, Lewin and others indicate, is actually more in tune with Clementsian ecology. Clements' climax communities, Chaoticians and other New Scientists would say, are examples of self-organising orders emerging from chaos.

One of the claims of those who support Chaos theory and complexity theory is that they give birth to views of nature that liberate nature's constituents. Thus we have people like E.C. White saying:

> The role ascribed to stochastic self-organisation in this view of natural history is an emancipatory one. ...The emergence of order out of chaos overcomes entropic degradation. Nature is both 'free' and 'progressive'.[47]

But if Chaos theory and Complexity theory are more linkable (or linked) to Clements than to Gleason, this claim for 'freedom' seems to come under some doubt since Clements superorganism tends to act as an imprisoning metaphysical framework as far as the species in a community are concerned. What's more, if Clements' superorganism is an example of the New Sciences of self-organisation, chaos and complexity at work, then so must Odum's ecosystems and Lovelock's Gaia since all three (Clements' climax communities, Odum's Ecosystems and Lovelock's Gaia) have been shown to be philosophically related. Therefore, far from having a role of emancipation as depicted in the above passage by White, self-organisation theory and Chaos theory would impose a metaphysical schema of systems determinism, over-generalisation, and uniformity upon nature's supposedly 'free' members. Just as Clementsianism, Gaianism and systems ecology imprison living beings within an over-lording holism which goes on to trap them in hierarchical levels and tyrannous niches, so the New Sciences of Chaos theory and self-organisation may do the same.

Perhaps it should be noted that the complex systems theory of the New Sciences does not break from uniting the natural world in a way that is any different from 'normal' systems theory. All of its basic assumptions are still there: systems, unity, holism, as well as levels and hierarchy, as we shall soon see.

Another possible relationship between ecology and the New Sciences is introduced by historian Donald Worster. Worster explains that the continued growth of Gleasonian-type ecological views in science--which talk of individualism and chaos in ecological settings–mirrors the accent on chaos that the New Sciences supposedly emphasise and that both of these disciplines together (i.e. Gleasonian ecology and Chaos theory) mirror the prevalence of individualistic and chaotic situations in modern day liberal industrial society. "It is hard to exaggerate..." says Worster:

> ...how far industrialism has gone in breaking down all the old notions of stability, community and order. Our entire world-view has been transformed profoundly by this force.[48]

Even if one accepts the thesis that general social conditions, and their popular appreciation, may give rise to cogent worldviews (such as the one of chaos and individualism that Worster speaks) it is very difficult to accept Worster's periodisation of this relationship that he draws between Gleasonian ecology, Chaos theory and modern day industrialism. If modern industrialism and liberalism have in some way imposed a worldview upon people by their manufacture of social chaos and individualism, then why did not Gleasonian-type individualistic/chaotic ecological ideas and the metaphysics of Chaos theory emerge at the beginning of the Nineteenth century rather than in the mid/late Twentieth Century as actually happened?

In any case, and as has been outlined above, Worster's asserted link between Chaos theory and Gleasonianism is hardly justified since Chaos theory and Gleasonianism give rise to conflicting interpretations of the reality of ecological situations. Chaos theory says that although things might appear to be chaotic, this is but a prelude to observable orders such as climax communities and stable ecosystems, whereas Gleasonianism would indicate that ecological settings are in a state of chaos without any hope of order emerging. Having made these points, however, there maybe a link between Chaos theory and the appreciation of chaos in modern liberal societies and this link is explored in the next section.

Chaos, Self-Organisation and Capitalism.

'Chaos frees the universe!'. This is the claim of the those who champion the New Sciences[49]. The chaos in Chaos theory supposedly frees the constituents of the universe from being trapped in the prison of Newtonian determinism:

> In classical Newtonian mechanics, once the initial conditions and the force laws are given, everything is calculable for ever before and after. The system is governed completely by the laws of mechanics and of conservation of energy. It is totally determined. It has no freedom.[50]

Now, however, we are said to be entering a new paradigm that says:

> the cosmos is a self-organising and quasi-sacred process that is developing greater complexity and greater freedom.[51]

Under the metaphysical framework of the New Sciences we now live in a universe that has the potential to evolve. The traditional scientific outlook which holds that the universe is a pre-designed clockwork mechanism slowly grinding towards thermodynamic decay is an outlook that is said to be, itself, grinding towards decay in current culture. Instead, we have an organic model of the universe; a universe that evolves towards greater and greater complexity and a universe whose constituents are not predetermined in their behaviour.

According to many New Scientists this new emerging paradigm, of evolution instead of decay and indeterminacy instead of determinism, is an answer to the age-old problem of reconciling order and freedom since the various constituents of the universe are free from predetermined actions of a central designer but the actions they do undertake nevertheless contribute to an overall order. In this scheme of things:

> The two forces that we have always placed in opposition to one another--freedom and order--turn out to be partners generating viable, well-ordered, autonomous systems.[52]

If we are to treat the 'order from chaos' idea with the respect that many New Scientists think it deserves then we must remember that the order from chaos idea is not just a phenomena of physics or of biology but is a phenomena associated with all sorts of things. Capra for instance believes that the science of 'order from chaos':

> makes it possible to begin to understand biological, social, cultural and cosmic evolution in terms of the same pattern of systems dynamics, even though the different kinds of evolution involve different mechanisms.[53]

In Ayres' *Information, Energy and Progress*, in Davies' *The Cosmic Blueprint*, in Merry's *Coping With Uncertainty*, in Capra's *The Web of Life*, in Mainzer's *Thinking In Complexity* and in a host of other books, this universal process of order from chaos is laid out chapter by chapter in the different disciplines and subjects of study that Capra mentions in the above quote. For instance, after explaining how the universe emerged as order from chaos, and after explaining how the planet Earth and its geological and biological components ordered themselves from chaos, Ayres goes on to devote consecutive chapters of his book explaining how order from chaos gave rise to advanced life, ecological systems, social systems, and economic systems. A similar pattern unfolds in the other books listed above as well as many other books and articles.[54]

Now, the attempt by Chaoticians to free the universe are not novel. Nor are their self-declared aims to reconcile order, chaos and freedom. The historian Otto Mayr points out that a number of thinkers from the Liberal tradition in the Eighteenth century attempted the same thing. According to Mayr, thinkers like Samuel Clarke, David Hume, Christian August Crusius, Leonard Euler, Colin Maclaurin and others were keen on finding ways to construct a worldview which supported their political commitment to liberalism and classical economics.[55]

As part of the necessity to legitimise and promote the ideas of liberalism, what liberals really liked were metaphors derived from situations in which there was no obvious authoritative or autocratic control of individual entities (and their activities) yet the situation as a whole did not actually slide into chaotic disorder. Mayr suggests that one of the places that they could find such situations were within the machines of the age.

For the likes of Clarke and Hume, their was often an intense philosophical interest in the burgeoning technology which went hand in hand with a parallel interest in the philosophy of the new politics of the day. To reconcile order and chaos, to enable the freedom of individuals without assuming a chaotic social universe, Clarke and Hume had to be careful in their use of machines as political and social metaphors. Certainly, as Mayr explains, the metaphors of the old age would not do. Clocks--which indicated an autocratic watchmaker and an intervening winder--were only ever used in political and philosophical writings as metaphors of authority and autocracy.

But other newly developed machines were of use, notably the so called self-correcting or self-regulating machines which didn't need the constant attendance of a minder. It is these self-regulating machines that were used as metaphors for liberalism since they were capable of regulating themselves without intervention. Says Mayr:

> The notion of the self-regulating system, applicable to the most diverse fields, splendidly matched the needs of the liberal concept of order and was well on its way to broad popular acceptance in Britain by the mid-Eighteenth century.[56]

In his book, Mayr cites some examples of the different metaphorical use of the clock compared to self-regulating machines to support his argument. He then goes on to boldly claim that the importance of the self-regulating machine metaphor is obvious if we consider the contemporaneous rise of liberalism in the British Isles compared to its much later arrival on the European scene.

The most well established and well-known self-regulating machine of the time was, of course, the steam engine. Here was a device that could adjust to internal and external variations and make appropriate corrections without the need of constant supervision. At this time the steam engine was hardly put forward by James Watt or any other of its inventors as a masterly example of self-regulation and the inherent compatibility of order and freedom. But for those with a philosophical bent it was a tangible example of the operation of self-regulation.

Mayr goes on to describe Adam Smith's 'Invisible Hand' as the apex of the philosophical tussles about order and freedom. This, Mayr makes out, can thereby be regarded as the concretisation of the liberal worldview. He states:

> The grand conclusion of the interdependent, almost symbiotic evolution of the concepts of self-regulation and the liberal system of economics was reached in Adam Smith's classic book 'The Wealth of Nations'.[57]

In this book, Adam Smith believed he:

> explained how the self-balancing mechanism of the free-market regulated the economy better than the most benevolent, omniscient central authority.[58]

For Mayr, Smith's Invisible Hand "is nothing other than the quality of self-regulation."[59] Mayr believes that Adam Smith's analysis of feedback in the economy is presented:

> in language so clear and is conceptualised so generally that it can be translated into the notation of modern systems theory without the need for any additional modification.[60]

Just as in ecological theory, where self-regulation has been a concept that enables the manufacture of theories and metaphysics involving natural unity, so too does self-regulation in economics give rise to notions of unity:

> It was the notion of the Invisible Hand that enabled Smith to develop the first comprehensive theory of the economy as an interrelated system.[61]

It is not only Mayr that delights in the contemporaneous development of self-regulating machines and the rise of liberalist economics. Advocates of the New Sciences have also been known to endorse the relationship too. Krohn and Mainzer, for example, state:

> The specific regulating mechanism of the feedback that Smith had propounded may be found in the new power machines that had a great impact on further development...The temporal simultaneity in articulating the idea of self organisation in nature, economy and technology is astounding.[62]

> From a qualitative point of view, Adam Smith's model of a free market can already be explained by self-organisation.[63]

Now, whether Adam Smith is in anyway a progenitor of Self-organisation theory--as Mayr, Krohn or Mainzer might suggest--is not really as important for us as the modern day renderings of his theories as being compatible to the self-organisation ideas contained within the New Sciences.

The economic equilibrium of the Invisible Hand has passed through various theories and ideas since Smith. It is, however, still a central concept in much economic analysis. Often historians of science have linked this concept to the appropriation and misappropriation of physical science theories and Newtonian metaphors.[64] I wouldn't want to debate the importance of these approaches but in the early years of the Twenty First century, liberalist/capitalist economics may be more likely to find metaphysical and scientific support from systems theory and systems versions of biology than it is from Newtonian physics.

The Ecosystem and the Economy: are they 'Birds of a Feather'?

Adam Smith's metaphysical musings about political economics were not merely specific intellectual tools for use in economic analysis. His notion of self-regulating order emerging from unconscious and chaotic actions was a full blown metaphysical outlook. Adam Smith, like Edward Goldsmith, Paul Davies and Fritjof Capra today, applied his ideas broadly; to economics, demographics, social theory, justice theory and the social differentiation of labour.[65]

Like Smith in the Eighteenth Century, those who try to reconcile order and chaos nowadays espouse them as near universal constants. As well as being applicable to societies, machines, the cosmos, the universal principle of order from chaos is also held by New Science fans to be observable in that bastion of ecological and environmental existence; the ecosystem. According to those who champion the 'order from chaos' idea the ecosystem is yet another example of a self-regulating order that emerges out of jumbled chaos. Davies, Capra and Birch, for instance adhere to this view.

Some adherents of the New Science bent, however, give the 'economy as ecosystem' analogy an extra twist. It is not that just any economy is analogous to just any ecosystem, it is that fully matured and self-regulating ecosystems show the same properties of organisation, process and complexification as capitalist free-market economies. For instance, in his book about the 'New Evolutionary Paradigm' of self regulation and complexity, Robert Ayres says:

> There can be no question that the operation of a money-based, free competitive market generates a kind of coherence, or long range order, somewhat analogous to so called co-operative phenomena.[66]

Ayres goes on to conclude that the Modernist foe of capitalism is not of this type of self-regulating complexity since socialism requires administration by intervention and planning by an overlooking orderer. Mainzer, another New Scientist, also reflects this attitude when he first celebrates the Free Market of Smithian economics as an example of self-regulation only to go on to announce that planned economies do not possess the natural regulatory benefits of market economies:

> Smith underlined that the good or bad intentions of individuals are not essential. In contrast to a centralised economical system, the equilibrium of supply and demand is not directed by a program-controlled central processor, but is the effect of an Invisible Hand, i.e. nothing but the non-linear interactions of consumers and producers.[67]

The reversible metaphor between the ecosystem and free-market economies is brought into new emphasis by Michael Rothschild in his book entitled *Bionomics: the Inevitability of Capitalism.* In this book Rothschild believes he shows that all of life is:

> a self-organising phenomenon. From the interplay of hormones in the body to the expansions and contractions of the great Arctic caribou herds, nature's intricately linked feedback loops automatically maintain a delicate, robust balance. Markets perform the same function in the economy. Without central planning, buyers and sellers constantly adjust to changing prices for commodities, capital and labour. A flexible economic order emerges spontaneously from the chaos of the free market.[68]

Believing that "a capitalist economy can best be comprehended as a living ecosystem."[69] Rothschild then goes on to declare that because ecosystems do not need to be planned to function well, and because economies are like ecosystems if allowed to run free, then planning in economies will only ever lead to trouble:

> Capitalism was not planned. Like life on Earth it did not need to be. Capitalism just happened, and it will keep on happening. Quite spontaneously. Capitalism flourishes whenever it is not suppressed, because it is a naturally occurring phenomenon. It is the way human society organises itself for survival in a world of limited resources.[70]

Rothschild's book is an exercise in the naturalisation of capitalism, and he admits as much, stating that he regards capitalism "as an inevitable, natural state of human economic affairs. Being for or against a natural phenomenon is a waste of time and mental energy.[71]

Rothschild's worldview is of intellectual as well as political importance since it shows some of the metaphorical processes at work when biological metaphors enter non-biological study. What we see from looking at Rothschild's work is that he first assumes an economics based on biology (something which is largely done for him by numerous predecessors) then he applies this economics back to biology (by using 'bionomics'--'the study of the economic relations of organisms and their environment') and then he brings this bionomics back to the social realm and sees in it economics again. This is a prime example of a reprojective spiral metaphor that has consistently accompanied nature/society analogies--both within and without the unity of nature idea.[72] It shows in Rothschild's case, also, that there is no original grounded base from whence the original metaphor flowed.

Now, environmental advocates of the New Sciences would probably like to disown anything that Rothschild might have to say but he is only deriving his worldview from the same place as they are; i.e. the liberal philosophy of order and freedom, especially as interpreted by the self-organisation ideas of the New Sciences and mixed with popular ecological concepts. Although ecologically minded New Science advocates like Fritjof Capra, Frederic Ferre and Charles Birch often rely on the New Sciences to bolster the scientific credibility of their political aims and environmental evaluations, they would probably be aghast to learn of the links that New Science has with the ideological metaphysics of modern industrial capitalism.

The links, both present and potential, might be further exposed if we note that it seems as though one of New Science's loudest advocates tacitly shares Rothschild's commitment to the metaphysics of capitalism. Paul Davies wrote in an article for *21st Century* magazine[73] that the model of the worlds economy as a ship captained by a steersmen should be dropped for a model of the economy as a self-organising ecosystem. This is also something Davies was prepared to declare as a fellow of the World Economic Forum.[74]

A little more explicit about the connection between the New Sciences and neo-classical economic ideas is the complexity theorist Klaus Mainzer who, as already noted, states that: "From a qualitative point of view, Adam Smith's model of a free market can already be explained by self-organisation".

In contrast to Paul Davies and Klaus Mainzer (who might be regarded as closet philosophical capitalists that passively use the ideas of New Sciences to promote a worldview amicable to their political commitments) there are some committed philosophical capitalists who actively tie the ideas of the New Sciences to the operation of Free Market economies. Rothschild does this with gusto but we may also note in this regard how More[75] enlists theories of self-generating order that come from the New Sciences to 'prove' the symbiosis of order and chaos in capitalism.

In presenting order and chaos as symbiotic, in the way that More and many others do, precise definitions of order and chaos have to be arranged. For More, order equates to economic equilibrium and progress, and chaos equates to individualistic actions that are not interfered with from above

Accompanying the above 'intellectual' Chaos theory fans, there are some practitioners of capitalism who themselves have cottoned on to the rhetorical use of self-organisation. For instance, a captain of industry that shares Davies' view about the equivalency of ecosystems and economies is Tachi Kiuchi, managing director of the Mitsubishi Electric Corporation in the United States. Kiuchi believes that to really get going and gain maximum efficiency players in the economy have to learn from (self-regulating) ecosystems; adjusting and adapting, as the components of ecosystems have, to feedback processes, niche competition, and new environmental conditions.[76]

More theoretical than Kiuchi are the economists Parker and Stacey who refer to the connection between the New Sciences and liberal capitalism by indicating that Chaos theory shows us that economies are best able to adapt and self-organise (and thus provide for the needs of the economy's members) when they are unplanned and unregulated:

> Chaos theory adds an important dimension to the study of economics. It helps explain why...an economic system is required which encourages adaptability... and...enterprise. Competitive markets have an important role to play in this process. Unlike planned systems, they provide for spontaneous adaptation.[77]

Another much more famous celebrant of capitalist economics who also shared these views is Frederick Hayek. Hayek, an intellectual figurehead for the right-wing Libertarian movement, was one of the foremost champions of Twentieth century capitalism. Hayek always embraced the order from chaos worldview of liberalism and often set about to detail some of its workings.[78] Spontaneous orders, as he called them, are the results of the actions of individual entities but not of conscious planning by these entities. Humans act, so thought Hayek, individually and rationally upon decentralised information flows, most notably price levels, to contribute to a spontaneous economic order.

If Hayek saw the emergence of order from chaos as a pretty-well universal phenomenon in nature and society then might there not be a link between this viewpoint of his and his celebration of capitalism. According to one of his intellectual historians, Robert Kley, there is such a link.[79] We can then develop Kley's ideas to show how Hayek's work is clearly, at the metaphysical level, intellectually attached to the New Sciences and Self-organisation theory.

Hayek, like many others, cast his philosophy of self-organisation rather widely. Where Adam Smith saw order from chaos in economics and social theory, where John von Neumann saw it in machines and cell biology, where Davies today sees it in ecosystems and solar systems, Hayek also saw spontaneous order in a myriad of places; from crystals and organisms to animal societies and galaxies. His favourite place to observe the machinations of self-ordering complexity, however, was, of course, the Free Market:

> Spontaneous social orders are 'the result of human actions but not of human design', the unintended consequence of the independent decisions and actions of many individuals.[80]

Although he used his own language and terminology when explicating the formation of spontaneous orders Hayek did see that the processes he had identified were compatible to those invented by the cyberneticists within their developing self-organisation ideas. As Kley indicates, for Hayek, models of order from chaos in economic situations could be explained in terms of information flow, feedback mechanisms and self-generation. Writing in the 1970s--a time that is often interpreted as the historical cusp between the systems theory of cybernetics and the complex systems theory of the New Sciences--Hayek became interested in complexity and organisation. He wrote:

> with spontaneous orders...their unplanned emergence must arouse some curiosity and warrants the establishment of a distinct body of theory.[81]

This distinct body of theory would, no doubt, be claimed as being the preserve and pursuit of the New Sciences by *its* scholars. Where Hayek in the 1970s looked forward to a "theory of complex phenomena". Davies announced in the 1980s that:

> There exists something like a law of complexity. But the study of complexity is still very much in its infancy. The hope is that by studying complex systems in many different disciplines, new universal principles will be discovered that might cast light on the way that complexity grows with time.[82]

Therefore it is probably safe to conclude that Hayek would have approved of the emerging disciplines of self-organisation theory and complexity theory that New Scientists have become so intensely fond of.

Following Kley's work on Hayek it is also possible to see the parallel between Hayek's excitement in his discovery of complexity with the current excitement exhibited by New Scientists over their discovery of the same phenomena. Hayek wanted to supersede the simple causal physics of Newton with a more complex science. Something that Davies (and Capra and most other New Scientists) also talk about a lot. Kley summarises these ideas of Hayek by saying:

> To bring out the features of complex phenomena Hayek contrasts them with 'simple phenomena'. The number of elements constituting the order of simple phenomenon is small. The orderly structure of its elements is the effect of a few one-way causal relations, and these relations are captured by the basic laws of physics. Finally, its environment does not influence the formation of a simple order. A complex order on the other hand, consists of a large number of elements and is the result of manifold exchange processes among the elements and between them and their surroundings.[83]

Without knowing it Hayek, in expressing the above ideas, could have been setting up the program of research that has now become complexity theory.

One well-known modern-day ecologist (who would call himself not a 'systems ecologist' after systems theory but a 'complexity ecologist', after complexity theory) has noted the relevance of Hayek's ideas for his own ecological studies. Donald deAngelis states that it is reasonable to think of a complex ecosystem in the same way that Hayek thought of the market. Just as the market exists as a spontaneous order from the chaotic actions of individuals, so the ecosystem exists as the self-ordered product of the species and populations within it. DeAngelis makes clear that he is not the only 'complex ecologist' to proffer such views:

> This view of the ecosystem as arising from the selfish interactions of species populations has been emphasised by some ecosystem theorists.[84]

Although Hayek's philosophical affinity to the metaphysics of the New Sciences might be considered scary enough for those contemplating the social and ecological relevance of the New Sciences, there are some even more disturbing intellectual developments for the likes of Capra. In an interesting use of the self-organisation concept and the ecosystems thought of systems ecology, the Libertarian scholar Barry Maley has claimed that the science of ecosystems justifies *not* protecting ecosystems. This writer believes that since the ecosystem is a prime example of a self-regulating order, the best thing we could do to save ecosystems is to run our economies like them. Thus, Maley goes on to suggest, environmental protection must be left to the workings of the market, the only economic system that obeys the self-ordering processes of ecosystems, and ecosystems will actually be protected. If we interfere with the machinery of the Market, suggests Maley, by implementing artificial regulatory regimes such as public reserves, environmental regulation, wildlife centres/refuges, and eco-taxes, then the economy will collapse from being self-ordering and the ecosystems will not survive:

> It is the preoccupation with achieving ends quickly, by fiat rather than adaptive process, which characterises political thinking and command and control makeshift.[85]

From this perspective, the ecosystem is held to act in accordance with its own processes only when the social equivalent of those processes--that is unfettered capitalism--is allowed to act. While environmentalists such as Capra, Birch, Goldsmith and Merchant, might regard unfettered capitalism as a major factor in the destruction of ecosystems[86] they unwittingly contribute to a worldview that suggests that environmentalism should be based on not directly protecting ecosystems.

Are Self-Regulating Entities as Free as They are Made Out to Be?

Having been inspired by the complex systems theory and its relevance to global order, Robert Artigiani writes:

> Only highly autonomous individuals, empowered to monitor and regulate their own experience, can collect and communicate the information necessary to preserve social structure in technologically advanced environments.[87]

If we are convinced, as is Artigiani, that the New Sciences are compatible with liberal capitalism and if we know that order emerges from chaos and that chaos gives rise to order, then all we need to do to gain order in the economy is to allow chaos to reign. Remember that for Hayekian influenced New Scientists such as More, Ayres, Rothschild, Mainzer and Maley 'order' is defined as economic equilibrium plus progress and 'chaos' is defined as the uninterrupted actions of unregulated individuals. Therefore to achieve order we must allow chaos. Even though one might disagree with the proposition that such order emerges out of such chaos it might reasonably be assumed that actually allowing for this type of chaos would be a pretty easy thing to do; you just let people do what they want to do. However, for the person that has continuously inspired the thinking of the above named metaphysical capitalists, Frederick Hayek, it was never that easy, as Kley relates:

> Spontaneous order, he [Hayek] claims, arises out of the general observance of certain behavioural rules and the individual adjustment to local circumstances. As he puts it elsewhere: 'the formation of spontaneous orders is the result of their elements following certain rules in their responses to their immediate environment'...'The individual responses to particular circumstances will result in an over-all order only if the individuals obey such rules as will produce an order.' Even a very limited similarity in their behaviour may be sufficient if the rules which they obey are sufficient to produce an order.[88]

So here it seems we are only assured of order if the chaos that gives rise to it is somehow law-abiding. Individual agents of chaos are not absolutely free non-determined agents whose actions give rise to order, as is presented by More, Ayres, Parker, Stacey, Artigiani and Rothschild, they are 'rule obeying units' whose struggle for order submits them to rule observance.

This contradiction ('freedom vs role-observance') seems to follow Chaos theory wherever it might be applied. For instance, Wheatley has tried to impress upon her readers the relevance of the New Sciences and the order from chaos idea in the disciplines of organisation and management. The management or organisational system, she says:

> has infinite possibilities, wandering wherever it pleases, sampling new configurations of itself. But its wandering and experimentation respect a boundary.[89]

Wheatley also says:

> If we allow autonomy at the local level, letting individuals or units be directed in their decision by guideposts for organisational self-reference, we can achieve coherence and continuity.[90]

'Guideposts' and 'boundaries' are hardly terms reminiscent of the all-embracing freedom that Davies, Ayres, Rothschild and others have spoken about with regards chaos. They may be softer words than the 'control' used by Bertalanffy, Wiener and Lovelock when they talk of their systems but they amount to the same thing. Indeed; 'boundary observance' and 'guide-posted decision-making' are the rhetorical manifestation of a theory that attempts to align itself with freedom by merely softening words like 'control' and 'rules'. This is patently the case if we examine the arbitrary nature of the terms 'guidepost' and 'boundary' in Wheatley's case. The guideposts and boundaries of Wheatley are not those democratically arrived at via worker votes but are those ascribed by the profit endeavours of the organisation to which the (supposedly autonomous) agents belong. No revolutionary progress in organisation studies has been made by inserting the New Sciences into management theory because the 'boundaries' and 'guideposts'--once 'instructions' and 'task-settings'--still come from the organisation's hierarchically privileged (higher status) members.

We might also note that in Hayek's case, the rule-observance that individuals must obey to find overall economic order are not matters of irrefutable fact accepted by each and every member of an economy but specific rules that emerge out of power-invested members in that economy. That the rules of the economy seem to be somehow skewed to the advantage of those making them exposes the socially-constructed nature of the rules. The chaotic individualism that is claimed to exist in economies is thus exposed as a confidence trick. Each and every player in the economy or in the organisation must play within the rules and adapt themselves to them if they are to succeed and/or survive. Individuals can hardly operate in some independent manner as Wheatley and Artigiani describe.

Thus an important issue never confronted in the New Sciences when examining 'order from chaos' in social and economic settings is that of the nature of the boundaries and guideposts. There is the possibility that they are just as enslaving and imprisoning as the centralised control whose necessity Hayek and Rothschild and Wheatley think they have demolished.

So much for Chaos theory in the social world but what does it do in the natural world? When Davies speaks of freedom in post-Newtonian cosmology, it is a freedom that also respects boundaries. When Capra talks about the indeterminism of ecosystems, it is an indeterminism that is guided by rules. And when populations of organisms are freed from the bonds of mechanist thinking, as Charles Birch asserts they should be, they only achieve that freedom within certain limits. One of those boundaries, and a persistent theme in unitarianism of any sort, is the boundary of hierarchy.

The Reliance upon Hierarchy within Self-Regulating Systems

David Abram declares that:

> if...we assume that matter is animate (or 'self-organising') from the get-go, then hierarchies vanish, and we are left with a diversely differentiated field of animate beings, each of which has its own gifts relative to others.[91]

But is this so? Another idea that is apparent in the New Sciences and especially that of the science of self-organisation is hierarchy. In every type of self-organising system the system is said to work by use of hierarchical structuring:

> The tendency of living systems to form multileveled structures whose levels differ in their complexity is all pervasive throughout nature and has to be seen as a basic principle of self-organisation.[92]

A key criterion of this type of systems thinking, according to Capra:

> is the ability to shift one's attention back and forth between systems levels. Throughout the living world, we find systems nesting within other systems, and by applying the same concepts to different systems levels--e.g. the concept of stress to an organism, city, or an economy--we often gain important insights.[93]

Of course, Capra, and all other New Scientists for that matter, include a whole host of entities under the title of living systems: cells, organismal bodies, ecosystems, Gaia, nations, societies. About the universality of hierarchy, Rollo says:

> The widespread existence of hierarchies is not trivial but probably represents a necessary conformation for organisations of great complexity.[94]

According to systems thinkers like Wiener, hierarchies allow quick processing of information, rapid decision making, and they ensure the efficiency of the system as a whole. These thoughts emerge from complex systems theorists too as can be observed in the writings of self-organisationists such as Ayres, Capra, Mingers and Merry.

When detailing the relationship between hierarchy and self-organisation Ayres graphically draws some models of particular socio-political structures. He contrasts simple hierarchical structures like kingdoms, armies and the communist party, with the feedback hierarchy of the United States tripartite governmental system. The U.S. hierarchical political system is thought, by Ayres, as a self-organising hierarchy, much the same as a self-organising cell or ecosystem. For Ayres, the ballot box is the crucial feedback process that serves as the link between different hierarchical levels.[95] Joseph, an advocate of Gaianism, does this too when he happily compares Gaian feedback with:

> One of the greatest and most complex negative feedback systems operating in the world today is the American system of checks and balances, as set up in the U.S constitution.[96]

Herein lies a telling point about the application of self-regulation from natural realms to social realms. Whether or not the analogies are at all accurate, self-organisation is being used as a justification of the rightness of both Western social and economic traditions (in particular Western capitalism and Western democracy) and the rightness of hierarchy within these traditions. A more bourgeois, conservative and ethnocentric philosophy of nature would be hard to find. Not only is Western democracy and Western capitalism sanctioned by natural laws but so is hierarchy, posited now as a natural necessity for the efficient running of complex systems like the State.

Fractals, Blobs and Boundaries

There is within Chaos theory a discipline known as 'fractal geometry'. Fractal geometry is of interest to us due to its relation to reductionism and hierarchical organisation. This would be considered a bit odd by New Scientists since fractal geometry is of interest to them due to its supposed holistic, chaotic and complex character.

Fractal geometry had its day in the sun with the widely circulated Mandelbrot blobs of the late 1980s and early 1990s that were seen adorning all manner of professional and popular texts.[97] These blobs are characterised as being self-organising, complex and chaotic. Their relationship to the unity of nature has so impressed Fritjof Capra that he placed an example of one on the cover of his book

The Web of Life as though it were a magical graphic whose mathematics underpinned the unity of all nature.

What impresses New Scientists so much is the way Mandelbrot blobs exhibit what is called 'self-similarity':

> The most striking property of these 'fractal' shapes is that their characteristic patterns are found repeatedly as descending scales, so that their parts, at any scale are similar to the whole.[98]

Numerous examples of this self-similarity in the physical world are given by those who are impressed by it. For instance, as Zimmerman and Capra explain:

> there is a universal pattern in the relation between large and small-scale features. The best known example of this relation is that of coastlines, which repeat the same ragged features at every scale of measurement.[99]

> Mandelbrot illustrates the property of 'self-similarity' by breaking a piece out of a cauliflower. He repeats this demonstration by dividing the part further, taking out another piece, which again looks like a very small cauliflower. Thus every part looks like the whole vegetable. The shape of the whole is similar to itself at all levels of scale.[100]

Capra carries on in this vein to declare:

> There are many other examples of self-similarity in nature. Rocks on mountains look like small mountains, branches of lightning, or borders of clouds, repeat the same pattern again and again; coastlines divide into smaller and smaller portions, each showing similar arrangements of beaches and headlands. Photographs of a river delta, the ramifications of a tree, or the repeated branchings of blood vessels may show patterns of such striking similarity that we are unable to tell which is which.[101]

This self-similarity is also taken to be evident in biological organisation; cells; organisms; ecosystems, etc, exhibit the same patterns at different scales. Each level is a repetition of the level above it and below it since these levels, too, are self-organising and self-regulating unities. Thus, for the New Scientists fractal geometry graphically and mathematically proves the concrete existence of the levels of biological organisation. Indeed the notion of a hierarchy of levels of biological organisation is a manifestation of self-similarity.

Wheatley, in her attempts to have the New Sciences guide the field of organisational management, speaks fondly of fractals. She indicates that a fractal organisation is one in which self- "similar behaviours show up at every level in the organisation because those behaviours were patterned into organising principles from the very start."[102] Wheatley attempts to tie fractals in with the autonomy of individuals and the order from chaos idea by saying:

> The potent force that shapes behaviour in these fractal organisations, as in all systems, is the combination of simply expressed expectations of acceptable behaviour and the freedom available to individuals to assert themselves in non-deterministic ways. Fractal organisations, though they may have never heard the word fractal, have learned to trust in natural organising phenomena.[103]

If Jean-Francois Lyotard had read and then re-worded Wheatley's above statement, he'd have probably said something like:

> The decisions do not have to respect the individual's aspirations: the aspirations have to aspire to the decisions, or at least to their effects. Administrative procedures should make individuals 'want' what the system needs in order to perform well.[104]

Again, it seems the important thing, and what is always left undefined by New Science advocates, is the nature of the acceptable behaviours and the nature of freedom. If an individual is free to act within acceptable boundaries there is hardly any recourse for that individual if he or she believes these boundaries are not sufficiently open to fully enable his or her freedom.

This argument does not apply just to Wheatley's organisations but to geometrical fractals generally. For instance when Capra says:

> another important link between Chaos theory and fractal geometry is the shift from quantity to quality. As we have seen, it is impossible to predict the values of the variables of a chaotic system at a particular time, but we can predict the qualitative features of the system's behaviour.[105]

We see, here, that bounds and boundaries still affect the scope of what is measured. An organism operating via the theories of the New Sciences can thus wander around within its role but it can never leave it, and an ecosystem can fluctuate according to some measured parameter but it can never become unstable.

Although Capra and others have emphasized the qualitative graphic character--as opposed to the precise numerical quantity--of fractal blobs, fractal science is still an example of old science, not anything new. Instead of reducing living things to genes or matter or energy, New Scientists are saying that all living things have a fractal pattern which equates to the essence of their existence. Discovering this fractal pattern, they say, via geometric science will some how reveal to us to the elementary character of various living and non-living phenomena. Thus if fractal geometry succeeds to become a generally followed scientific programme it will be a programme grounded in reductionism since all phenomena in the universe will be reduced to the essences contained within their fractal blobs.

Self-Organisation and God

As well as offering supportive evidence for the naturalness and efficiency of the Free Market and for the existence of levels of biological organisation, a few New Scientists have held that Chaos and self-organisation in the universe to be manifestations of God. Davies is one who leads this charge:

> that the universe has organised its own self-awareness is for me powerful evidence that there is something happening behind it all. The impression of design is overwhelming. Science maybe able to explain all the processes whereby the universe evolves its own destiny, but that still leaves room for there to be a meaning behind existence.[106]

Both Davies and Capra paraphrase Reich Jantsch to summarise this state of affairs: 'God is not the creator, but the mind of the universe.'[107]

What is more, God is now--once more--fond of humans since we humans "have been written into the laws of nature in a deep and, I believe, meaningful way."[108] In this view:

> the deity is, of course, neither male nor female, nor manifest in any personal form, but represents the self-organising dynamics of the entire cosmos.[109]

This situation strongly parallels that identified by Robert Young within Victorian science where God changed from an entity-being (that made the harmony in nature and society that humans observed) to a deity identified with self-acting natural laws. At this time "Science did not replace God. God became identified with the laws of nature."[110] As much has happened in the work of the New Sciences when Davies states:

> The word God means so many different things to different people that I am loathe to use it. When I do, it is in the sense of the rational ground that underpins physical reality. Used in this way, God is not a person, but a timeless abstract principle that implies something like meaning or purpose behind physical existence.[111]

For Davies, this abstract principle is 'self-organisation'. Hence we come to a conclusion which Davies would, himself, honour: God is self-organisation.

The equation: self-organisation equals God has, in itself, all sorts of connotations apart from providing evidence for the existence of God. For one thing, it suggests scientists as the priests of the universe searching for the ultimate truths just as the clergy have done in centuries past. This undoubtedly adds to the authority of scientists when prescriptions are made for social change towards a supposedly more environmental way of being.

Another implication that flows from the desire to acknowledge the Godliness of self-organisation is the danger that we stand of deifying all those things that have so far been said to operate in a self-organising fashion. Goldsmith and Lovelock suggest we should actually do this in the case of Gaia but, more alarming, people like Rothschild, Ayres and Hayek say we should do this for capitalist Free Market economies. Free Markets thus become not only essential natural processes--as promoted by Rothschild and other capitalist New Science advocates--they also become sacred processes. The Free Market is thus the social and economic manifestation of the timeless abstract principle that Davies holds as God.[112] If this is true then the Free Market might soon be classed as a sacred untouchable divine entity worthy of reverence. The Free Market is not only natural, essential and unchanging but also omniscient, holy and transcendental. If this is so then a fairly obvious implication is that the Free Market should be treated with respect and that it should not be tampered with.[113]

The 'Self' in Self-Regulation

If we take note of the arguments presented in previous chapters and go on to disagree with the New Scientists and the systems theorists and announce that there is not any sustained self-regulation in ecosystems, and that there is no natural process in ecology which goes towards maintaining a definite spatio-temporal structure like the ecosystem, then we might soon arrive at a more general conclusion with regards to self-regulation in ecology. This conclusion says that the ecosystem is not a 'self' (which by common agreement we would take to mean a united entity) but merely a non-uniform mixture of separate entities that together are hardly even maintaining their own collective state of existence (since they are forever changing). This is the conclusion of Gleasonian ecology, which suggests that, at least in the realms of ecology, the 'self' in 'self-regulation' is an idealized abstraction.

This doubting of the self-hood of ecosystems does not necessarily mean we have to doubt the self in all the other supposedly self-regulating phenomena of the universe but it does direct suspicion towards some of them. Free Market ideologues, for example, cannot rely on the economy running like an ecosystem because the grounded object from which they draw their metaphor is not a coherently identifiable self undergoing self-regulation but a tossing and turning collection of disunited creatures.

Suppose we have successfully cracked and fractured the self-regulating/self-organising character of the ecosystem into non-existence, what does this mean for other levels of biological organisation that have, according to New Scientists, self-organised themselves from lower self-organised levels? It means that we might still be able to say that cells self-organise into organisms and that organisms self-organise into populations. After this, however, we are in trouble, for populations do not seem to self-organise into ecosystems so much as amass together to become associations of disparate individuals. So what we find is that the grand hierarchy (of levels of biological organisation) is fractured and disrupted at one point. Lower levels have seemingly self-organised into higher and higher wholes until, above the population level, we find that everything is a bit of a mess. Similarly, the biosphere--which Lovelock, Goldsmith, and other unitarian thinkers believe is the highest level of biological organisation--is not an entity that organised itself from the lower level of agglomerated self-organised ecosystems. Instead it is an entity that--if self-organised at all--did so without the help of ecosystems.

New Scientists, and unitarians in general, are often given over to an interpretation whereby what applies to one level of biological organisation applies to them all (this is usually reinforced, as has been noted, by the idea of self-similarity). The ecosystem concept is held to be such a strong link in this process of self-similarity and hierarchisation that it has often served as the referent from which to make new metaphors and analogies from. However, if the ecosystem level is destroyed then unitarians might wonder about the capacity for other levels to be named 'selves' and thereby for these levels to be elaborated in terms of the concepts of self-organisation and self-regulation.

Summary and Conclusions

Chaos theory believes it sanctions freedom. Chaoticians and other New Scientists thus value freedom, you might think. This may be so but just as likely is the idea that New Scientists know freedom has enormous political and philosophical clout and they attach themselves to its ideals by positing some (dubious) association between freedom and indeterminacy.

Chaos theorists also value order. Without it their beloved science cannot see or do anything. What they need to do, then, in order to maximise their rhetorical appeal to various contemporary discourses, is to entangle both freedom and order together.

Economic liberalists and capitalists also value freedom and order. The freedom of the individual and the order of the Free Market economy. Order, they say, can only exist with such freedom, and freedom can only exist if there is such order. Again capitalists ideologues can maximise their rhetorical appeal by somehow entangling order and freedom together into one.

What is interesting is that both New Scientists and liberal capitalists do this entangling in a very similar way, and what is more, they both appeal to the same historical and intellectual parentage when doing it--so much so that the language of the 'order from chaos' idea represented in Eighteenth Century liberal philosophical ideas is, according to Mayr, understood by both modern day liberal capitalists and modern day New Scientists. It is also apparent that the natural examples of order from chaos which are outlined by New Scientists (like Paul Davies) can help liberal capitalists further their cause (to naturalise and deify economic examples of order from chaos). Also, the examples of self-organisation outlined by liberalists (like Hayek) can be utilised by New Scientists to help further their cause (to arrive at universally true principles of the operation of the cosmos).

This is not just a possibility; it is actually happening and can be observed in the writings of people like Ayres, Maley, Mainzer and Rothschild who not only possess the desire to present universal principles but also have an obvious pro-capitalist agenda.

While this chapter might seem to advocate getting rid of the ecosystem concept (or, at least, some of the intellectual security it enjoys) because of the ease to which it becomes an ideological tool, previous chapters have indicated that we might get rid of the ecosystem concept because it fosters bad ecological practice. All these chapters together suggest also that the reality of the self-regulating ecosystem should be doubted. That the New Scientists can postulate self-organisation in ecology so readily where Gleasonian ecologists can only doubt it, is unwittingly explained by an advocate of universal self-organisation, Max More:

> The abstractness of the S.O.'s [self-organisations] make them particularly difficult for the untutored mind to recognise. You can't simply look at an S.O. and spot it. You need to apply a theory.[114]

Although More is a firm supporter of the existence of self-organisation in economics, physics and biology (and especially in the Free Market) he nevertheless knocks the nail right on the head with his admission of the abstract nature of self-organisation. In the case of self-organisation theory, it may only be a short step from admitting that self-organisation is an abstraction to admitting that a self-organising phenomenon can be conjured up anywhere and in virtually any situation because the theory in question posits such simple processes and lives within such vague parameters. First, you take the worldview of unity, second, add a preoccupation with systems, third, add feedback loops, fourth, characterise the process using arrows or some other dynamic symbolism and there you have it; a self-organising phenomena. Thus it is not hard for any theorist to postulate self-organisation all over the place.

A way to destroy this line of abstract universalism is to derail the integrity of the 'self' in self-organisation. We might note, for instance, that ecosystems succumb to such a derailment for it can be suggested using a wide variety of counter-evidence and argument that they are unregulated and un-regulating collections of entities that hardly fulfill what it is to be a self. Economies, too, might succumb to this derailment since they are hardly unitarian phenomena but consist of all sorts of conflicting and varying processes, behaviours, events and agents.

New Scientists, and environmental writers who support the New Sciences, would of course advise against getting rid of the concept of self-organisation since it has shown itself to be useful in explaining and exploring various natural wonders through a new perspective which suggest that the world is alive, free and organic. Self-organisation, and the other theories of the New Sciences, are thus useful in ecopolitical discourse since they dramatically express the aliveness of the Earth and many of the Earth's processes. This aliveness, New Science fans believe, surely asks more of humanity when it comes to protecting the Earth and its environmental processes than the mechanical worldview of Newton. Thus, the New Sciences are useful for garnering an ecopolitical and ecophilosophical view of the living world.

However, this is not the only thing the New Sciences are useful for. They may also be useful in resurrecting scientists as the priests and arbiters of the nature of reality, useful in delegitimising any economics not based on Western Free Market principles, useful in reinvigorating systems science as a universal discipline, useful in legitimising hierarchy in modern society, and useful in limiting environmental evaluation to specific scientific and scientistic management practices. These uses can be considered to be contradictory to the programme of environmental consciousness that many environmental scholars and philosophers of nature are seeking to forge.

Notes for Chapter 5 (Uniting the Ecosystem and the Economy)

1. A. Naess, "Deep Ecology", in C. Merchant, ed, *Key Concepts in Critical Theory: Ecology* (NJ: Humanities Press, 1994) p.122.
2. R.A. Peterson, eds, *System, Change and Conflict* (NY: Free Press, 1967) p.117.

3. E.D. Schulze and H. Zwolfer, eds, *Potentials and Limitations of Ecosystem Analysis* (Berlin: Springer Verlag, 1987) p.1. We need not rely upon the testimony of systems experts and unitarians, however, when finding links between unity and systems. Any dictionary may be consulted to give similar indications of the link. The *Burlington Universal Dictionary*, for example, defines a system as being "an assemblage of things forming a connected whole". Similarly, unity is defined in a similar way: the "oneness" that comes from a "harmonious association" within a "connected whole". Such a connection doesn't necessarily designate a direct equation between those who hold that the world is a system and those who hold it acts in unity but it does at least bring to the fore the idea that there is the possibility of a socially agreed philosophical relationship between the two since the language used by supposedly disinterested dictionary editors to define 'system' is common to the language used to describe a 'unity'.
4. For more about this with specific relevance to systems theory see R. Lilienfeld, *The Rise of Systems Theory: An Ideological Analysis* (Krieger, 1988).
5. Incidentally, when the activity of one thing influences the activity of another thing in a positive way; i.e. when there is a reinforcement and encouragement of a certain activity, we get positive feedback. This type of feedback, though, is generally regarded as being pretty useless since the system then spirals out of control and into nonexistence. Such would be the case, for instance, if the sensor in a refrigeration unit told the motor to turn on whenever the temperature in the compartment got too cold.
6. For an introduction to Hutchinson's work see: R. P. McIntosh, *The Background to Ecology* (NY: Cambridge University Press, 1985); L.A. Real and J.A. Brown, *Foundations of Ecology: Classic Papers with Commentaries* (Chicago: University of Chicago Press, 1991).
7. Hutchinson quoted in C.L. Kwa, "Radiation Ecology, Systems Ecology, and the Management of the Environment" in M. Shortland, ed, *Science and Nature*, (BSHS Monograph, 1993), p.223.
8. E. Goldsmith, *The Way: An Ecological Worldview* (NY: Shambhala, 1993) p262.
9. J. Hagen, *An Entangled Bank: The Origins of the Ecosystem Concept* (New Brunswick: Rutgers University Press, 1992), p71.
10. F. Capra, *The Web of Life*, p.54
11. *ibid.* We might like to compare this statement of Capra's with the tales of some biologists who went to these conferences and who actually found them pointless (see Hagen, *An Entangled Bank*). Incidentally, within Capra's hagiographic treatment of systems science he thinks, for some reason, that we should pay attention to early systems theory because it emerged from people with fast brains. About von Neumann, for example he says:

> "John von Neumann was...a mathematical genius, he had written a classic treatise on quantum theory, was the originator of the theory of games, and became world famous as the inventor of the digital computer. Von Neumann had a powerful memory and his mind worked with enormous speed. It was said of him that he could understand the essence of a mathematical problem almost instantly and that he would analyse any problem, mathematical or practical, so

clearly and exhaustively that no further discussion was necessary." (quote from Capra, *The Web of Life*, p.118).

12. *ibid* p.54.
13. Goldsmith, *The Way* p.130.
14. J. Lovelock, *The Ages of Gaia* (NY: WW. Norton & Co, 1988).
15. Capra, The *Web of Life*, p.63.
16. Kwa, "Radiation Ecology, Systems Ecology, and the Management of the Environment", p.247.
17. *ibid*, p.248
18. D.R. Biggins, "Biology and Ideology", *Science Education*, 60, no. 4, pp567-578. (1976). Twenty years later a similar, though less obvious, influence upon ecological metaphors via management concepts can be found in Fritjof Capra's *The Web of Life*. Where, for twenty odd years the 'system' was Capra's prime metaphorical device to explain the nature of all existence (see F. Capra, *The Tao of Physics*, Boston: Shambala, 1975; F. Capra. *The Turning Point*, NY: Simon & Schuster, 1982; F. Capra, *Uncommon Wisdom*, NY: Simon & Schuster, 1988; and Capra, *The Web of Life*) now he is beginning to swap the 'system' for the 'network'. On occasion we can see Capra use 'network' in exactly the same way he has used 'system'. Where Capra's systems were 'dynamic, integrated unities of parts', so his networks are 'dynamic, integrated unities of parts'. Thus Capra now says: "during the second half of the century the network concept has been the key to recent scientific understanding not only of ecosystems but the very nature of life" (Capra, *The Web of Life*, p.35.)--where he would have said the exact same thing for systems in his earlier publications. This could easily lead one to ponder--in the vein of Biggins' 1976 assertion--whether or not Capra's new emphasis on networks are the result of the continued growth of networks and network analysis in both management and computing and especially with regards to the Internet. (It should be acknowledged that this transition from the system metaphor to the network metaphor is not complete within Capra's writing and the majority of his explications within his latest works still afford reference to systems).
19. Goldsmith, *The Way*, p.108.
20. See: E.P. Odum, *Fundamentals of Ecology*, (Philadelphia: Saunders, 1971).
21. Kwa, "Radiation Ecology, Systems Ecology, and the Management of the Environment", p.249.
22. See S.I. Auerbach, S.I. "Foreword: George M. Van Dyne - A Reminiscence" in B.C. Patten & S.E. Jorgensen, eds, *Complex Ecology* (NY: Prentice-Hall, 1995) p.xxvii-xxx. In this reminiscence, Auerbach also had this to say about George van Dyne's management style:

    > "To say that he was somewhat dogmatic is a bit of an understatement. When he was manager of the program, his forcefulness more than occasionally resulted in controversy with other scientists, especially those with similar traits. He firmly believed that his approach to ecosystem research in the grasslands was the appropriate model or paradigm for all the biome programs".

23. F. Geyer, "The Challenge of Sociocybernetics", *Paper Presented to the World Congress of Sociology*, Bielefeld, July 18-24, 1994.
24. For example, we may quote the systems theorist Rosen who said:

    "the developing family of ideas and concepts which fall roughly under the rubric of systems theory amounts to a profound revolution in science--a revolution which will transform human thought as deeply as did the earlier ones of Galileo and Newton" (quote from R. Rosen, "Review of Trends in General Systems Theory" *Science*, 177, (1972) :508.

25. For instance consider the proclamations about complex systems theory of Krohn *et al* who are: "observing a present day scientific revolution", and Kuppers who says we are "experiencing a singular phenomenon in the development of science, for which I can think of no better description than fundamental" and Arthur Fabel who believes that "the current ferment in science is potentially more than another Copernican revolution." These quotes come from W. Krohn, *et al*, "Self-Organisation--The Convergence of Ideas: An Introduction", in W. Krohn *et al*, eds, *Self-Organisation: Portrait of a Scientific Revolution* (Dordrecht, Kluwer, 1991) p.1; B-O. Kuppers, "On a Fundamental Shift in the Natural Sciences", in W. Krohn *et al*, eds, *Self-Organisation: Portrait of a Scientific Revolution* (Dordrecht: Kluwer, 1991,) p.51; A.J. Fabel, "Environmental Ethics and the Question of Cosmic Purpose", *Environmental Ethics*, 16 (1994) p.304.
26. See H. Maturana and F. Varela, *Autopoiesis and Cognition*, Dordrecht: Kluwer, 1980); E. Jantsch, *The Self-Organising Universe* (Oxford: Pergamon Press, 1980); I Prigogine, and I. Stengers, *Order Out of Chaos*, NY: Bantom Books, 1984).
27. Capra, *The Web of Life*, p.x.
28. P. Davies, P "The Future of God", *Sydney Morning Herald*, Dec 21st, (1996) p.6.
29. Capra, *The Web of Life*, p.112.
30. See N.K. Hayles, *Chaos Bound: Orderly Disorder in Contemporary Literature and Science* (Ithaca: Cornell University Press, 1990); N.K. Hayles, ed, *Chaos and Order: Complex Dynamics in Literature and Science*; (Chicago: Chicago University Press, 1991).
31. Quotes taken from: P. Davies, *The Cosmic Blueprint* (Harmondsworth: Penguin, 1987) p.51.; Capra, *The Web of Life*, p.122.;
32. Quotes taken from: M. Wheatley, *Leadership and the New Science: Learning About Organisation from an Orderly Universe*, (San Francisco: Berret-Koehler Publishers, 1992) p.11.; U. Merry, *Coping With Uncertainty: Insights from the New Sciences of Chaos*, Self-Organisation and Complexity (Westport: Praeger, 1995) p.13.
33. *ibid*, p.13.
34. F. Capra, "Systems Theory and the New Paradigm", in C. Merchant, ed, *Key Concepts Critical Theory: Ecology* (NJ: Humanities Press, 1994) p.340.
35. Capra, *The Web of Life*, p.301. The same point of view can be found in systems biologists by perusing S, Camazine *et al*, eds, *Self-Organisation in Biological Systems* (Princeton University Press: Princeton, 2001).
36. R. Lewin, "All for One, One for All" *New Scientist*, 152 (14 Dec 1996), p28.

37. Davies, *The Cosmic Blueprint*, p.132.
38. See: Capra, *The Web of Life*, p.100; F. Pearce, Gaia's Guardian Angel, *Resurgence*, 186 (1998). (Repr: online at: www.gn.apc.org/resurgence).
39. J. Cohen and I. Stewart, *The Collapse of Chaos*, (London: Viking, 1994) p.387.
40. Merry, *Coping With Uncertainty*, p.30.
41. *ibid*, p.61.
42. Fraser Clarke quoted in: R. Wright, "Art and Science in Chaos: Contested Readings of Scientific Visualization", in R. Robertson *et al*, eds, *FutureNatural* (London: Routledge, 1996) p.227.
43. Capra, "Systems Theory and the New Paradigm", p.340.
44. Capra, *The Web of Life*, p.7.
45. The precepts of Deep Ecology are expressed in the well-known Deep Ecology eight point platform. It reads like this: 1) The well-being and flourishing of human and non-human life on Earth have value in themselves. These values are independent of the usefulness of the non-human world for human purposes. 2) Richness and diversity of life forms contribute to the realisation of these values and are also values in themselves. 3) Humans have no right to reduce this richness and diversity except to satisfy basic needs. 4) The flourishing of human life and cultures is compatible with a substantial decrease of the human population. The flourishing of non-human life requires such a decrease. 5) Present human interference with the non-human world is excessive, and the situation is rapidly worsening. 6) Policies must therefore be changed. These policies affect basic economic, technological, and ideological structures. The resulting state of affairs will be deeply different from the present. 7) The ideological change is mainly that of appreciating life quality (dwelling in situations of inherent value) rather than adhering to an increasingly higher standard of living. There will be a profound awareness of the difference between big and great. 8) Those who subscribe to the foregoing points have an obligation directly or indirectly to try to implement the necessary changes. This platform is announced in various publications, including: B. Devall and G. Sessions, *Deep Ecology: Living As if the Earth Really Mattered* (Layton: Gibbs M. Smith, 1995) and A. Naess, *Ecology, Community, Lifestyle* (Cambridge: Cambridge University Press, 1989; A. Naess, "Deep Ecology", in C. Merchant, ed, *Key Concepts in Critical Theory: Ecology*, Humanities Press, NJ, 120-124. It should be noted that Capra does not talk about any of these things in *The Web of Life*, despite saying his philosophy is a cosmological manifesto for Deep Ecology
46. See for example Devall and Sessions, *Deep Ecology*; Naess, *Ecology, Community, Lifestyle*; and Naess, "Deep Ecology".
47. S.K. White, *Political Theory and Postmodernism* (Cambridge: Cambridge University Press 1991) p.264.
48. D. Worster, *The Ecology of Order and Chaos*, in S.J. Armstrong and R.C. Botzler, eds, *Environmental Ethics: Divergence and Convergence*, NY: McGraw-Hill, 1993, 39-48
49. 'Chaos Frees the Universe!' was the title of an article in *New Scientist* by Paul Davies, See P. Davies. "Chaos Frees the Universe" *New Scientist*, 128 (6th Oct. 1996), 36-39.
50. C. Birch, *On Purpose* (Sydney: UNSW Press 1991) p.ix
51. M.E. Zimmerman, *Contesting Earth's Future: Radical Ecology and PostModernity* (Berkeley: University of California Press, 1994) p.14.

52. Wheatley, *Leadership and the New Science*, p.13.
53. Capra, *The Turning Point*, p.310.
54. A short list of such books and articles might include: J. Gleick, *Chaos* (London: Penguin, 1987); F.D. Peat, *The Philosopher's Stone: Chaos, Synchronicity and the Hidden Order of the World* (NY: Bantam Books, 1987), C. Dyke, *The Evolutionary Dynamics of Complex Systems: A Study in Biosocial Complexity* (NY: Oxford University Press, 1988), J. Briggs and D. Peat, *Turbulent Mirror: An Illustrated Guide to Chaos Theory and the Science of Wholeness* (NY: Harper and Row, 1991); F.E. Yates, *Self-Organising Systems: The Emergence of Order* (NY: Plenum Press, 1989); G. Roth and H. Schwegler "Self-Organisation, Emergent Properties and the Unity of the World", in W. Krohn *et al*, eds, *Self-Organisation: Portrait of a Scientific Revolution*, (Dordrecht: Kluwer Academic, 1990), pp.36-50; S. Kauffman, *The Origins of Order*, (NY: Oxford University Press, 1993); G. Scott, *Time, Rhythm and Chaos In the New Dialogue with Nature* (Ames: Iowa State University Press, 1991); P. Davies, *The Mind of God*, (London: Penguin, 1993); S.J. Goener, "Reconciling Physics and the Order-Producing Universe: Evolutionary Competence and the New Vision of the Second Law", *World Futures*, 36, (1993) :167-179; M.M. Waldrop, *Complexity: The Emerging Science at the Edge of Order and Chaos* (London: Viking, 1993); R. Abraham, *Chaos, Gaia, Eros: A Chaos Pioneer Uncovers the Three Great Streams of History*, (San Francisco: Harper, 1994); J. Cohen and I. Stewart, *The Collapse of Chaos* (London: Viking, 1994); R.J. Russell, *et al, Chaos and Complexity: Scientific Perspectives on Divine Action* (Vatican City: Specola Vationa, 1990), S. Kauffman, *At Home in the Universe* (NY: Oxford University Press, 1995), J. Mingers, *Self-Producing Systems (*NY: Plenum, 1995); G.M. Hall, *The Ingenious Mind of Nature: Deciphering the Patterns of Man, Society and the Universe*, (NY: Plenum Trade, 1997); T. Bossomaier and D. Green, *Patterns in the Sand: Computers, Complexity and Life* (Sydney: Allen and Unwin, 1998), S. Camazine, *et al*, eds, *Self-Organisation in Biological Systems* (Princeton: Princeton University Press, 2001).
55. O. Mayr, *Authority, Liberty and Automatic Machinery in Early Modern Europe*, (Baltimore: Johns Hopkins University Press, 1986).
56. *ibid*, p.139.
57. *ibid*, p.172.
58. *ibid*, p.165.
59. *ibid*, p.175.
60. *ibid*, p.176.
61. K.I. Vaughn, *Austrian Economics in America: the Migration of a Tradition* (Cambridge: Cambridge University Press, 1987) p.168.
62. W. Krohn, *et al*, "Self-Organisation--The Convergence of Ideas: An Introduction", p.2.
63. K. Mainzer, *Thinking in Complexity: The Complex Dynamics of Matter, Mind, and Mankind*, Berlin: Springer-Verlag, p.11.
64. For example, see D. McClosky, *The Rhetoric of Economics* (Madison: University of Wisconsin Press, 1985); P. Mirowski, *Against Mechanism: Protecting Economics from Science* (NJ: Rowman & Littlefield, 1988). and P. Mirowski, *More Heat Than Light:*

Economics as Social Physics, Physics as Nature's Economics, (Cambridge: Cambridge University Press, 1989);

65. See N. Barry, "The Tradition of Spontaneous Order", in *Literature of Liberty*, Sept 1982, 45-67; and J.Z. Muller, (1993*) Adam Smith in His Time and Ours: Designing the Decent Society* (NY: Free Press, NY, 1993).
66. R.U. Ayres, *Information, Entropy and Progress* (Woodbury: AIP Press, 1994) p.134.
67. Mainzer, *Thinking in Complexity*, p.11
68. M. Rothschild, *Bionomics: The Inevitability of Capitalism* (NY: Henry Holt, 1990) p.xiv.
69. Rothschild, *Bionomics*, p. xi.
70. Rothschild, *Bionomics*, p.xi. Incidentally Rothschild obligingly outlines those things that upset the natural balance of the natural economy (in the vein of Lovelock's identification of those things that upset the natural balance of Gaia). These things are nationalisation of industries, income taxes, profit taxes, looking after the unemployed, and helping bankrupts. An obvious alignment between Rothschild economic thinking and rationalist economics appears to be present. His views on this topics are expanded in a list of articles published on the website: http://www.bionomics.org/
71. Rothschild, *Bionomics*, p.xv.
72. For explorations of reprojective spiral narratives in the biological sciences external to the unity of nature idea, see: R. Young, "The Historiographic and Ideological Contexts of the 19th Century Debates on Man's Place in Nature", in I. Teich and R. Young, eds, *Changing Perspectives in the History of Science: Essays in Honour of Joseph Needham*, (London: Heinemann, 1974) pp.334-438; S. Schweber, "Darwin and the Political Economists", *Journal of the History of Biology*, 13, 195-289 (1980); R. Young, R. *Darwin's Metaphor* (Cambridge: Cambridge University Press, 1980); T. Porter, "Natural Science and Social theory", in R.C. Olby *et al* ,eds, *Companion to the History of Modern Science* (London: Routledge, 1990) pp.1024-1043; I. Cohen, I. *Interactions: Some Contact Between The Natural and Social Sciences* (Cambridge: MIT Press 1994); P. Mirowski, ed, *Natural Images in Economic Thought*, (Cambridge Cambridge University Press, 1994).
73. P. Davies, "A Vision of Science in the 21st Century", *21st Century*, 12, 102-103 (1994).
74. P. Davies, "Ants in the Machine", *Sydney Morning Herald*, Oct 17th 1998, p.6s
75. See M. More, "Order Without Orders", *Extropy*, 3, p.21-32. (1991). Also see: J. Lesourne, *The Economics of Order and Disorder: The Market as Organizer and Creator* (Oxford: Clarendon, 1992).
76. R. Neville, "The Future Isn't What It Used To Be", *Good Weekend* May 16th, 1998, pp.16-22.
77. D. Parker and R. Stacey, *Chaos, Management and Economics*, (Sydney: Centre for Independent Studies, 1995) p.76.
78. F.A. Hayek, *Studies in Philosophy, Politics and Economics*, (Chicago: Chicago University Press, 1967).
79. R. Kley, *Hayek's Social and Political Thought* (Oxford: Clarendon, 1992).
80. Hayek quoted in Kley, Hayek's *Social and Political Thought*, p.102. As Kley explains, although Frederick Hayek saw spontaneous orders in all sorts of places, his favourite subject of examination was the economy. Order from chaos in this realm meant the economic equilibrium and economic progress emerged from the actions of chaotic

individuals as they played and acted in the Market Place, haphazardly buying and selling to push and pull supply and demand.

81. Quoted in Kley, *ibid*, p.38.
82. Davies, *The Cosmic Blueprint*, p.21.
83. Kley, *Hayek's Social and Political Thought*, p.41.
84. D.L. DeAngelis, "The Nature and Significance of Feedback in Ecosystems in B.C. Patten and S.E. Jorgensen, eds, *Complex Ecology*, (NJ: Prentice Hall, 1995), p.463.
85. B. Maley, *Ethics and Ecosystems* (Sydney: Centre for Independent Studies, 1994), p.92.
86. For instance see Devall and Sessions, *Deep Ecology*; A. McLaughlin, "Ecology, Capitalism and Socialism" *Socialism and Democracy*, 10, pp.69-102. (1990); J.M. Gowdy, "Further Problems with Neo-Classical Environmental Economics", *Environmental Ethics*, 16, pp.161-171(1994); D.R. Loy, "The Religion of the Market", *Journal of American Academy of Religion*, 65, pp.275-290 (1997).
87. R. Artigiani, "Post-Modernism and Social Evolution: An Inquiry", *World Futures*, 30, pp.149-161 (1991).
88. Kley, *Hayek's Social and Political Thought*, p.28.
89. Wheatley, *Leadership and the New Science*, p.123.
90. *ibid*, p.95.
91. D. Abram, Returning to Our Wild Animal Senses, *Wild Earth*, 7, no. 1, (1997): 10.
92. F. Capra, *The Turning Point* (NY: Simon & Schuster, 1982), p.303.
93. Capra, *The Web of Life*, p.36.
94. C.D. Rollo, *Phenotypes: Their Epigenetics, Ecology and Evolution* (London: Chapman and Hall, 1995) p.8.
95. Ayres, *Information, Entropy and Progress*.
96. L.E. Joseph, *Gaia: The Growth of an Idea* (NY: Arkana, 1990), p.114.
97. For examples of texts elucidating and promoting the virtues of fractal geometry with reference to Mandelbrot blobs, see: B.B. Mandelbrot, *The Fractal Geometry of Nature* (NY: W.H. Freeman, 1982) and all those references listed in note 54.
98. Zimmerman, *Contesting Earth's Future*, p.348.
99. *ibid.*
100. Capra, *The Web of Life*, p.125.
101. *ibid*, p.137.
102. Wheatley, *Leadership and the New Science*, p.132.
103. *ibid.*
104. Capra, *The Web of Life*, 137.
105. J-F. Lyotard, *The Postmodern Condition* (Manchester: Manchester University Press, 1984), p.62. Also, in reference to organizational management, see: D.M. Boje and R.F. Dennehy, *Managing in the Postmodern world: America's Revolution Against Exploitation* (Dubuque: Kendall/Hunt, 1993).
106. Davies, *The Cosmic Blueprint*, p.203.
107. See Capra, *The Turning Point*; Davies, *The Cosmic Blueprint*; and E. Jantsch, *The Self-Organising Universe*.
108. Davies, *The Mind of God*; p.196.
109. See Capra, *The Turning Point*.

110. R. Young, "Science, Ideology and Donna Harraway", *Science as Culture*, 3, 2, (1992) :165-206.
111. Davies, *The Mind of God*, p.21.
112. It might be noted that Paul Davies in the mid 1990s, was a grateful recipient of the $750,000 Templeton Prize from the Templeton Foundation for his popular science writings. This Prize was awarded to Davies for his 'original contributions to humanity's understanding of God' via self-organisation and complexity theory. Along with the Templeton Foundation's ongoing project to utilise science to discover God, the other major project of the foundation is to understand and celebrate the fundamental philosophy of 'Free Enterprise'. In light of the arguments in this section, it now appears that Davies might be considered to have contributed intellectually to this project as well.
113. Many critics of the Free Market would probably comment here that such deification of the Market has already permeated culture via the growth of economic rationalism and monetarism and the resurgence in 'neo-classical' liberal economics. If we acknowledge (in line with *Penguin Dictionary of Politics*) that Hayek, and his economic philosophy of 'self-regulating order emerging from chaos', was "largely ignored by economists and politicians for most of his career" (which spanned from the 1930s till his death in 1992) but "became extremely influential in the 1970s and 1980s, when his ideas found favour with the Reagan administration in the USA and, above all, with the Thatcherite wing of the Conservative Party" and we also note the contemporaneous rise of self-regulation as a scientific idea, then we can see just how much monetarism might be the practical result of the master narrative that is self-organisation. If the deified soul of economic rationalism is exposed to be 'self-organization' then the importance of self-organisation as the primary metaphysical commitment of 1980s and 1990s Western culture might be hinted at. While I like to think that this chapter has opened up the possibility of such an exposition; a deep exploration of the idea that 'self-organisation actually is *the* global cultural myth of this turn of the millenium period' must remain a project for future study. My purpose in this book is just to expose the mythic potential of self-organisation within environmentalism, science and the philosophy of nature.
114. More, "Order Without Orders", p.22.

PART C:

It's all a Postmodern Plot(lessness)!

Part C focuses in on a particular idea: that the unity of nature idea is a construction of various breeds of postmodernism. These breeds of postmodernism attempt to locate themselves in opposition to what they term Modernism. Part C, therefore, is devoted to an investigation of the postmodern attitudes that may lay within, or contribute to, the contemporary enunciation of the unity of nature. It examines the claim that if we postmodernise science--making it reject the Modernist ideals that stem from the onset of Modernity during the Renaissance and Enlightenment--then we can produce a 'Postmodern Science' that offers a post-mechanistic, post-reductionistic, post-atomistic, post-fragmentary worldview.

Such a project seems to sharply reflect the ideals of the New Paradigm whose historical narrative was outlined in Chapter One; and indeed the New Paradigmers and the Postmodern Science supporters are the same people much of the time. What can be found, however, is that Postmodern Science is also related to systems theory and the New Sciences. (By now the reader will have realised that the name of these theoretical nodes of thought often change but much of the underlying metaphysical themes are the same). If this is so (and if systems theory and the New Sciences can be shown to be replete with the Modernist predilections towards reductionism, mechanicism, hierarchy, and also universalist theory) then Postmodern Science might be so attached to Modernist ways of doing science that it actually is not postmodern at all.

After deconstructing Postmodern Science, Part C enters into a critique of one of the dichotomies which is being played out within the stories of the New Paradigm (as well as being played out within the New Sciences and Postmodern Science); that is the story of 'mechanicism versus organicism'. Postmodern Science supporters generally want to uphold the organicism versus mechanicism dichotomy; placing themselves distinctly within the organic dipole for this gives them what they believe is an inherently postmodern perspective on nature. My argument is that the mechanicism versus organicism division is hardly of the nature that Postmodern Scientists make out. Advancing oneself as an organicist, as opposed to a mechanist, is a rhetorical ploy which plays within the folds of Modernism (rather than postmodernism). A postmodern deconstruction of the organicism versus mechanicism dichotomy reveals that organicists and mechanicists are often espousing very similar worldviews.

The final chapter of this Part, and of the whole book, addresses the deconstructed remains of the unity of nature idea. From these remains, Chapter Eight seeks to construct the characteristics of what a more aggressively postmodernised environmental worldview might look like, especially with regards to the area of environmental concern most immediate to the minds of many unitarians: the terrestrial ecological communities of the world. If environmentalists do need to find themselves a narrative about the world's ecology, it is suggested in this chapter how they might find one through a creative disintegration of ecological unity to give rise to a worldview of Otherness.

CHAPTER 6

What Is This Thing Called Postmodern Science?

Introduction

In an effort to critique the scientistic ideals that developed strongly throughout the Nineteenth century, scholars in the past couple of decades have come to formulate a particular term to summarily describe the dominant Western intellectual traditions and practices since c1850; this term is: 'Modernism'. Modernism stands out as the self-aware phase of a longer period of history often referred to as Modernity, a time when humanism took over from religious orthodoxy, when the capital classes began exerting control over the aristocracy, when industrialism was emerging, and the proletarianization of the rural and urban masses was accelerating. Where Modernity might be said to extend back beyond Newton and Descartes to the start of the Renaissance, Modernism is often given the starting date of 1846 when Charles Baudelaire defined it in relation to contemporary art[1]. In this regard, Modernism might be regarded as the conscious or thinking phase of Modernity.

Some of Modernism's oft-stated characteristics are faith in the emancipatory potential of mass democracy, faith in capitalism as an agent of economic evolution and freedom, faith in technology and industry to better the 'human condition', and the transfer of faith from religion to science and from God to Progress. The contemporary theoretical reactions against such new faiths, such as that of Karl Marx and William Morris, are also held to be an inherent aspect of Modernism; part of it's self-conscious reflection which just transposed faith from capitalist economics toward socialist or anarchist economics.

Modernism stands out (sometimes starkly, sometimes not) against what is believed to be 'postmodernism'. Some believe that postmodernism can be chronologised as that time from about the postwar period, especially since the 60s, until now. During this time new modes of capitalism, art, architecture, culture and communication have invaded the social world; an invasion sometimes dubbed 'postmodernity'. The new world of postmodernity required new theories of society, new theories of culture, of language and of art and politics that Modernism could not provide.

Postmodernism is held by postmodern scholars as being an inherently more suitable way to approach the study of culture, society, science and art since, unlike Modernism, it doesn't get caught up in scientific or industrial fetishism, it relinquishes the stifling ideology associated with a search for the ultimate Truth, it promotes an inherent celebration of diversity and difference, whilst simultaneously multiplying the stories and images and ideas within society.

Having heard about this critique of Modernism, many philosophers of nature have decided that this is what they were also critiquing. Thus influential environmental scholars like Donald Griffin, Charles Birch, Carolyn Merchant, David Bohm, Charlene Spretnak and Frederick Ferre have recently announced the rise of 'Postmodern Science'.[2] Out of the ashes of receding Modernism they have endeavoured to embark on firing up a Worldview that avoids the pitfalls of Modernist science, such as extreme reductionism, mechanicism and atomism.[3] According to the likes of Griffin and Merchant, these Modernist pitfalls have been brewing in Western culture since before the Enlightenment but reached a crescendo when scientists had fully bought into Modernism in the late Nineteenth century. Yet if these pitfalls can be ejected from science, then so much the better, they feel, for the environment since a new environmentally friendly worldview will be encouraged to take centre stage in the thought and culture of the Western world.

The point which is made in this chapter, however, is that Postmodern Science might well be so inculcated with Modernism (as it is defined by postmodern intellectuals and as it is defined by Postmodern Science itself) that it can hardly be called postmodern. One of the ways that this inculcation is made manifest is the close association between Postmodern Science and the systems theory of the New Sciences. This association infects Postmodern Science with various Modernist impulses so that the primary project of Postmodern Science--to create an environmentally friendly worldview--is undermined.

Postmodern Characteristics

As unitarians like Griffin, Merchant, Birch, Ferre, Bohm and Gare are so fond of identifying themselves with the term postmodernism, it is as well that we identify what postmodernism actually is. Before going on to detail the core theoretical premises of Postmodern Science, it is worth taking a look at the intellectual movement from which Postmodern Science takes its name. By doing this we can then go on to carefully analyse to what extent Postmodern Science is an expressions of postmodern sentiments.

However, when defining postmodernism we immediately come into problems. Postmodernism defies definition. For reasons which may become more obvious in the following section, postmodernists are loathe to define the movement they are involved with since such self-definition might totalise the identity of the movement towards a narrow, restrictive and objectifying synecdoche, or as Sassower puts it:

> providing a definition goes against the postmodern grain since it pretends to capture a moment that is too fleeting to catch.[4]

This penchant for ephemerality and this disdain for the restrictive notion of a singular identity are not just principles of postmodernist anti-definitionism, however, they permeate through the whole gamut of postmodern subjects of study. Postmodern sociologists decry the use of identity in social studies, for instance; choosing instead to focus on differences, diversity, and heterogeneity. In doing this they often believe that they have unearthed the ephemerality and changeable nature of social phenomena. The social world is hardly stable, say postmodernists, or even progressively changing in an evolutionary manner, nor does it operate in clearly observable cycles or any other observable pattern.

It could be noted that within this the perspective, all of life (sociological life or ecological life) is transient and indeterminate:

> Indeterminacy often follows from fragmentation--the postmodernist only disconnects; fragments are all he pretends to trust. His ultimate opprobrium is 'totalisation'--any synthesis whatever, social, epistemic, even poetic. Hence his preference for montage, collage.[5]

In this process of disconnecting and dis-aggregating the total whole, the category of 'Life' may itself be fractured; becoming ephemeral, diverse, fragmented, heterogeneous and decentralised. What emerges from the ashes of 'Life' are pluralised 'lives'; the category of Life succumbs to the overwhelming expression of different lives; lives which are too diverse to be able to be categorised by the one common abstract notion of 'Life'

These tendencies towards heterogeneity, difference and plurality, along with ephemerality and anti-identity, also serve as some of the core descriptions of postmodernism:

> pluralism...has become the irritable condition of postmodern discourse, consuming many pages of both critical and uncritical inquiry.
>
> The very foundation of postmodernity consists of viewing the world as a plurality of heterogeneous spaces and temporalities.
>
> Postmodernism lends itself to...heterogeneity without any critical ordering principle.[6]

The heterogeneous nature of language, society, geography and culture--and the transient nature of intellectual, social and artistic trends, truths and traditions--promotes heterogeneity and ephemerality as common postmodern characteristics over different subjects of study. As West would put it, postmodernism tends to:

> trash the monolithic and homogenous in the name of diversity, multiplicity and heterogeneity; to reject the abstract, general and universal in light of the concrete, specific and particular; and to historicise, contextualise and pluralise by highlighting the contingent, provisional, variable, tentative, shifting and changing.[7]

Along with difference and diversity, fracture and fragmentation are also often held to be symptoms of postmodern thought. Whether they are talking about the fracturing of the signified from the signifier, the fragmentation of social structures into cultural groupings or the breaking up of unilinear history into multiple histories, fracturing and fragmentation of some sort or another often lies within the studies of postmodernists.

Needless to say this advocacy of fracture, indeterminacy and pluralism has often been equated with intellectual decadence, nihilism, and impracticability by some writers.[8] It is acknowledged within postmodern circles that postmodernism:

> rejects epistemological assumptions, refutes methodological conventions, resists knowledge claims, obscures all versions of the truth, and dismisses policy recommendations.[9]

Because of this project, postmodernism is sometimes seen as intellectual chaos with little ability to address major intellectual or practical problems. Or as Rod Tester puts it:

from the point of view of certain modern boundaries postmodernity seems to be direction-less, blurred and lacking in rigour.[10]

Tester, however, goes on to reply on behalf of postmodernists that:

from the point of view of postmodernity, modernity is a prison of one sort or another towards which the only proper attitude is incredulity.[11]

This incredulity appears because Modernism marginalises what is held to be unusual, generalises rather than specifies, and congeals together things which should be allowed to fragment apart.

For many who have in some way utilised or sympathised with postmodernism, the broken nature of the social, cultural or geographical world is only a secondary result of postmodern analysis. The primary result is the unearthing and deconstruction of ideas that have hitherto been thought of as being part of fundamental knowledge. Although they sometimes try to deny that what they are doing is analytical, postmodernists nevertheless critically analyse the founding ideas of their subjects of study to see what previous students in the field have failed to see; that these founding ideas are merely assumptions. Founding ideas, otherwise known as foundations, are thought by many in the postmodern movement to be the core problem within Modernity and their deconstruction the core project of postmodernism. Foundations are said to be expressed most cogently through the enunciation (and re-enunciation) of metanarratives:

...in Lyotard's view, a metanarrative is meta in a very strong sense. It purports to be a privileged discourse capable of situating, characterising and evaluating all other discourses, but not itself infected by the historicity and contingency which render first-order discourses potentially distorted and in need of legitimisation.[12]

What Is This Thing Called Postmodern Science?

Although postmodernism might like to characterise science as the epitome of Modernism there is within professional and popular science circles an ongoing attempt to postmodernise science. Having been made aware of the postmodern critique of Modernism, some scientists and science writers who have been critical of the traditional scientific methods and principles have chosen to affiliate themselves with postmodernism. Instead of rejecting the ideas of science altogether, however, they opt for replacing Modern science with 'Postmodern Science'. Therefore, it may be said that instead of harbouring desires for a revolution against science, they want reform.

As in other intellectual fields that are involved with postmodernism, at the forefront of Postmodern Science there is an interest in metanarratives. Instead of totally abandoning all metanarratives and all foundational ideas, however, Postmodern Science supporters want to resurrect environmentally friendly and socially benevolent ones. For many 'Postmodern Scientists' this is the program of Postmodern Science; to resurrect more appropriate foundational ideas about the natural world.

Postmodern Science's penchant for the reconstruction of foundational ideas (rather than their total de(con)struction) in the wake of general postmodern and anti-science attacks in the last thirty years, leads Postmodern Scientists to label their peculiar brand of postmodernism 'Constructive Postmodern Science'. As Griffin points out in his book, and the series in which it is located[13], whereas most postmodernism has a penchant for breaking down bastions of Modernism in a destructive way:

> The postmodernism of this series can, by contrast, be called constructive or revisionary. It seeks to overcome the Modern worldview not by eliminating the possibility of worldviews as such, but by constructing a postmodern worldview through a revision of modern premises.[14]

Where postmodernists have generally been critical of the competence of science to reveal the truths of the natural world, and where postmodernists are often critical of the worldviews that science does reveal, Postmodern Scientists believe that science has now got something new to offer. Contemporary science, it is claimed, is now beginning to understand the universe in a whole new light; a light supposedly compatible with many postmodern concepts.

Introducing this new Postmodern Science, Arthur Fabel says there is "a new science, just now arising...with a conception of the cosmos as a self-organising genesis."[15]

This one comment of Fabel's is, even by itself, quite telling. As he explicitly refers to the 'new science' (i.e.: New Science) of 'self-organisation' as being Postmodern, Fabel seems to want to ally Postmodern Science with the self-organisation ideas of the New Sciences that were discussed in the last chapter. This connection is confirmed by the comments of numerous other Postmodern Scientists. For example, although Davies does not use the 'Postmodern Science' label, Gare labels Davies' New Science as Postmodern Science.[16] Paul Davies (and many who share Paul Davies' worldview) are pulled in to the circle of Postmodern Science by fans like Gare who enthusiastically classify Davies' scientific approach as Postmodern Science.

Charles Birch further strengthens the affinity between the complex systems theory of the New Sciences and Postmodern Science by repeating for Postmodern Science what he has for a long time said about systems theory:

> In a postmodern world, the new images are no longer mechanical: they are organic and ecological. The universe turns out to be less like a machine and more like a life.[17]

Birch, in *On Purpose*, then goes on to define more closely what Postmodern Science is in a section entitled: "Five Axioms for A Postmodern Worldview":

> The first axiom: Nature is organic and ecological...The postmodern worldview takes seriously the proposition that we live in a universe and not a multiverse. It is ecological through and through.[18]

Birch's second axiom of the 'Postmodern Worldview' declares that we must interpret lower levels of reality in terms of higher levels. His third declares that we should interpret the world in terms of monism as against a Nature/Humanity dualism. His fourth pleads us to adopt biocentrism and his fifth asks us to be post-disciplinary, as well as universalising:

> Knowledge is lost in a sea of beliefs from a multitude of disciplines. The general purpose of the modern university is lost amid the incoherent variety of special purposes that have accumulated within it. The outlook of the postmodern worldview is for fewer beliefs and more beliefs.[19]

We shall come back to the themes of Birch's fourth and fifth axioms latter on but for now we can identify his axioms 1-3 as programmatic statements that any Postmodern Science must look towards organicism, unitarianism and holism in some way.

For Hayles and Wiessert, that 'some way' towards organicism, unitarianism and holism is through Chaos theory and complexity theory. Hayles and Wiessert regard Chaos theory and complexity theory as examples of Postmodern Science because these theories supposedly reject the notion that physical matter is inanimate and particulate (composed of broken bits and pieces) and also because they negate the idea that physical processes are deterministic and isolated. Wiessert's and Hayles' readings of Chaos theory posit chaos as an indication of the complexity of order and not as the absence of orderliness and they suggest that such order can only be understood by new mathematical ideas which have yet to be fully elaborated.[20] Such new mathematics (which I suppose will be called 'Postmodern Mathematics') will liberate nature from determinism and go on to provide a dynamic view of matter.

Stephen Best would agree with most of this, stating: "Postmodern Science has three main branches of influence": thermodynamics, quantum mechanics, and Chaos theory, which altogether:

> emerge as a break from the mechanistic, objectivist and deterministic worldview of modern science.[21]

Best goes on to say that:

> Postmodern Science rejects the crippling dualistic outlook of modern thought; instead it sees nature, human beings, and the relation between human beings and nature in holistic terms.[22]

David Bohm's unitarian physics also displays predilections towards this use of the term postmodernism. For most of his long life, Bohm was without a label for his holistic physics but he found it just before he died: "I am proposing a postmodern physics that begins with the whole."[23] Similarly Frederic Ferre believes holistic ecology to be Postmodern Science, stating that Postmodern Science is a return to organic ecological ideas.[24]

According to Krippner the Modernist scientific worldview can be partially attributed to the unprecedented fragmentation, nihilism and destruction that we find in the world. He then goes on to say that as a corrective to this situation:

> 'postmodern' or 'holistic' thought hopes to preserve the virtues of the 'Modern' worldview while replacing its mechanistic and reductionistic assumptions with those that are more organic in nature.[25]

So Postmodern Science is the science of chaos and complexity, the science of self-organisation, the science of unity, organicism and holism. All this gives rise to the following conclusion. Postmodern Science is basically equivalent to the complex systems theories which in Chapter Three we called the New Sciences.

Sometimes this conclusion becomes clear when considering the prescribed political responsibility of both systems theory and Postmodern Science, respectively. Consider the similarities between Capra, an explicit advocate of 'systems theory' with those of Gare an explicit advocate of 'Postmodern Science'. In the same way that Capra, systems theorist, says that a new worldview must be forged for environmental reasons so Gare, Postmodern Scientist, says "what is required to address the environmental crisis" is a "postmodern cosmology"[26] And just as Capra labels Deep Ecologists the grass-roots organisers of the new paradigm, so Gare states: "Postmodern environmentalists, that is, the 'deep ecologists' and associated movements" are those who reject Modernist environmental destruction.[27]

Because of their very similar critiques of Modern science and Modernist living both Capra and Gare would, with little doubt, agree with Stephen Best's following comments on postmodernity and living in harmony with nature:

> in rejecting modernity as a historical era of wanton destruction, many advocates of postmodern science embrace the concept of postmodernity as a new historical period yet to be created--where human beings exist in harmonious relations with nature, each other, and their own selves.[28]

All this is to say that what is presented as holistic/unitarian thought in Chapter Two and as the New Sciences in Chapter Five is painted as Postmodern Science by those who, in this chapter, are called Postmodern Scientists. If this is so, then surely we have a contradiction emerging. Given the penchant for unity and generalisation and universalisation in the holistic sciences, systems theory and the New Sciences, and given the penchant for heterogeneity, fragmentation and difference in postmodernism, how can unitarians contemplate labeling holistic science or systems theory or the New Sciences as 'Postmodern'?

Postmodernism in (a) Word(s)

When investigating the legitimacy of Postmodern Scientists' claims to be postmodern, we can do so by aligning the commonly-accepted characteristics of postmodernism with the characteristics of Postmodern Science. To cover a lot of ground quickly this can be done by listing twenty or so one word characteristics of postmodernism and comparing them to the characteristics of Postmodern Science.[29] In this vein postmodernism can be characterised as being (or emphasising):

'Anti-scientific': (for example, as expressed by: Lyotard, Eagleton, Harvey, Best, Grenz and Grant)[30]

This feature is succinctly expressed by Grenz who declares: "postmodernism is the end of science."[31] Here Grenz expresses the idea that science is, in fact, the epitome of Modernism, and that the end of science as we know it will pave the way for postmodernism since it will involve the social revocation of scientific expertise, scientific management, and the end of power struggles based on knowing the scientifically-refereed truth. In light of this, the term 'Postmodern Science' ought to be considered an oxymoron. The fact that Postmodern Scientists can think up the term, and then espouse worldviews of unity from it, must also be seen as oxymoronic in the light of the postmodern critique of science. If Postmodern Science wishes to use science to erect a new philosophical worldview, then this sits uneasily with the claims of most other postmodernists who would declare that:

> science and philosophy must jettison their grandiose metaphysical claims and view themselves more modestly as just another set of narratives.[32]

On the count of postmodernism's strong anti-scientific character, then, it seems that Postmodern Science fails to be postmodern.

'Pessimism': (for example, as expressed by: Best, Kroker and Cook, Bauman and Grenz)[33]

According to Grenz a 'gnawing pessimism' pervades postmodernism. After being influenced by historian/philosopher Michel Foucault and others it now seems to many social scientists that revolutionary and reforming utopians are no longer able to validate their knowledge with references to the truth of things and the nature of nature since:

1) the knowledge (of nature and of society) that is said to be possessed by utopians/reformers/revolutionaries is not separable from their political aims, and

2) the politics and philosophies they espouse will be dangerous and oppressive to someone or other.

With regard to number 2, Zygmunt Bauman would want to draw our attention to the omnipresence of dilemma in the postmodern world. If the ideas associated with a utopian or revolutionary plan are ever realized in the name of 'liberation from oppression', genuine oppression of other ideas must necessarily follow for they cannot, too, be realized. In a (postmodern) world replete with the values of diversity and difference it seems as though any utopian that somehow has his or her utopia come to fruition will be totally incapable of sparing themselves:

> the bitter after-taste which comes unsolicited in the wake of decisions taken and fulfilled[34]

This dilemma-ridden angst within postmodernism is, however, a topic that is avoided in a major way in Modernism. Dilemmas between interests have always been smoothed out or smoothed over by abstractly converting the opposing parts into a whole whose ongoing well-being is served, maintained and delivered by the same processes that give rise to the dilemmas. For instance, the struggle between selfish economic competitors supposedly promotes innovation and enterprise to make the economic world a whole lot better for all, and in the natural world, the struggle for survival between species is heralded by Modernist scientists as driving evolution forward so that the living world as a whole becomes more complex and diverse. It seems, then, as though Modernism might be characterised as the era without any significant dilemma. Any dilemma, under Modernist rule, is solved simply by appealing to an abstract whole and the processes that form it, such as the Adam Smith's Invisible Hand, Spencer's 'survival of the fittest', Parson's social homeostasis, and Lovelock's Gaian self-regulation. From within such a Modernist metaphysical viewpoint it is hard to get gloomy about the dilemmas of interest which are cruelly maldistributed amongst the members of the world since in the end (and over the entirety of the world) the whole is held to be bettered by the presence of such dilemmas. However, if no abstract whole (like society, the economy or Gaia) exists then the processes that are said to regulate and maintain it are brought into question and the value of dilemma as a progressive and self-ordering force is nullified.

So where does this gnawing pessimism leave Postmodern Science? Firstly those within its ranks who harbour utopian ideals might be seen to be conflicting with postmodern thought. Utopias, including environmental utopias and intellectual utopias that the likes of Capra, Goldsmith, Birch and others describe, exist as narrow unitary visions with a certain specified state of social being and thus are not only programmatic challenges to a particular status quo (in how we view the world or how we should change it) but attempted constructions of idealistic unities which necessarily exclude those that can not or will not conform to them.

The pessimism of postmodernism is most ardent when considering the various technotopian ideals that spill out from technocratic Modernists. Far from enabling a better and better society, science and technology is only capable of producing self-denying nightmares of social poverty, injustice and environmental destruction. Science's ability to get a handle on understanding these problems is undermined by the Power-Knowledge relationships within science so that science (whether it is Modern science or Postmodern Science) ends up devoid of any ability to objectively convey to the planners of utopia and social betterment the grounded knowledge upon which value-neutral policy decisions can be made.

Postmodern Science, however, is not nearly so pessimistic. Postmodern Scientists are sure that new ways of conducting science, i.e. organically and holistically; and new ways of viewing the universe, i.e. organically and holistically, will infiltrate both technology policy and social policy to deliver a civilisation more at harmony with itself and its environment. Stephen Best reiterates this difference between postmodernism and Postmodern Science when he says:

> Although both postmodern science and social theory are critical of modern rationality, postmodern science tends to be far more optimistic about the value of science, technology and rationality than most postmodern theory.[35]

Whether Postmodern Science is correct in thinking that it can produce a socially and ecologically suitable scientific response to the world's woes is not the really the point here. The point is that Postmodern Science is incongruously positioned with regard to the attitude that most postmodernists would take towards the optimism of science and technological planning.

This incongruity between Postmodern Science and postmodernism can also be spotted in the following declaration coming from Paul Davies that "the new way of thinking about the world is more cheerful."[36] Davies here, as we noted in Chapter One, believes that the social acceptance of holism is directly related to collective despair promulgated by living in a reductionist universe. Although he does not offer any evidence to equate despair with reductionism, he nevertheless supposes the holism of Postmodern Science (he calls it New Science) to be relatively optimistic points of view. This, too, differentiates the optimism of Postmodern Science and its worldviews with the pessimism of much postmodern critique.

On the count of pessimism, then, it seems that Postmodern Science fails to be postmodern.

Anti-truth and Anti-realism (for example, as expressed by Baudrillard, Lyotard, Appignanesi and Lawson, Bauman, Rosenau, Zimmerman, Easthope and Scott)[37]

Postmodernists have often held Nietzsche to be a proto-postmodernist[38], so it is with one of his many aphorisms that I shall begin this sub-section. Nietzsche would say that 'truths are illusions that we have forgotten are illusions; they are metaphors that have become worn out'[39]. Postmodernists, too, have a particular (but not necessarily peculiar) take on both truth and reality; neither exist. That is to say truth and reality are no longer regarded as independent objective references but socially constructed contingencies decided upon temporarily by the social milieu of particular situations. "Any particular truth", so says Pauline Rosenau:

> is relevant or valid only to members of the group or community within which it is formulated. Knowledge, then, is relative to the community.[40]

Inspired by the likes of Bruno Latour[41], Rosenau goes on to summarise the postmodern view on reality as: "no external reality actually exists as the ultimate arbitrator."[42]

Far from being universal and independent of human activities, truth and reality are localised human inventions but, in addition to this, the concepts of Truth and Reality are themselves local inventions of the Modernist world which have been constructed and reconstructed by political and intellectual players to serve their own purposes. Truth and Reality in the Modern age have become such hugely important ideas because they exist as the socially-accepted final referee in the political games of all Modernists. Needless to say virtually all players in Modernity have inevitably described the referee of Truth and Reality as confirming their own claims. To play outside of Truth and Reality--to fail to utilise them in your knowledge claims during Modernity--has meant to immediately cast yourself off as superstitious or irrational, mad or deviant, pagan or uncivilised. Within postmodernity, however, political play without Truth and Reality has become acceptable.

For those postmodernists inspired by Foucault, Truth is intimately linked to Power:

> Truth does not consist of propositions that correctly 'mirror' or 'represent' an independent, pre-existing reality. Instead, what passes for 'objective' truth is a construction generated by power invested elites. [43]

For Foucault, Power, itself, is deconstructed and rearranged so that it does not resemble a pyramidal hierarchy as conceptualised within class, gender or race activism but instead becomes a pluralised entity reflecting a whole host of different and inter-relaying forms. Power thus operates in uniquely arranged local sites but always as part of the fabrication of truth:

> It is impossible to separate truth from power, and so there is no real possibility of any absolutely uncorrupted truth.[44]

If this is so, then environmentalists thus construct their truths (for instance; the unity of nature) as much around political aims and aspirations as do their anthropocentric industrialist foes (who may speak of the duality of Humanity and Nature, the superiority of Humanity, and the necessity to conquer Nature). Thus it can be said that:

> Postmodern theorists...insist that no one is innocent; every one...is concerned with defining truth as a way of acquiring and retaining power.[45]

For many critics of this facet of postmodernism, it seems that the debate over issues such as the existence of Truth and/or an independent reality are of interest only to intellectuals who, insulated from both the Truth and daily reality, never personally experience the violence, terror, and degradation prevalent in the contemporary world.[46]

From such a point of view, the postmodern engagement with the de(con)struction of all truths is also a disengagement from political fighting and social change altogether. Such disengagement with the 'real' issues of political and social life sometimes sounds very much like a conciliation to the enemy.

However, for postmodernists, there are ways around this. One can distrust the concepts of Truth and Reality without conciliating to the enemy. For instance:

> To be sure, finite beings have no access to absolute truth, and establishing finite truths is often a difficult, disputational, and politically charged process. Still, none of this prevents some historical, philosophical, or scientific texts from being more informative or illuminating.[47]

This might mean that whatever perspective we have of the world (whatever peculiar mixture of metaphors and stories and images we might hold to), it is always possible to augment them by dwelling upon commensurable metaphors and stories and images that have worked well enough when re-mixed, recombined and re-utilised in other ways. By this way of thinking, human 'Truth' never reflects natural 'Reality', but it may be a useful model to get things done. If this is so, then the operational mode of what was once seen to be True is now just acknowledged as a workable model to speak about, understand, and act within the world.

For some postmodernists, however, this approach might lie too close to the rhetoric of Modernist science to be palatable. Modern scientists still revel in and celebrate the dictums of science that state that science has no pretenses to finding the final word on all things but only offers up temporary solutions to intellectual and practical problems until better ones come along. *Nullius in verba* reads the Royal Society's motto; 'No man's word shall be final'. The problem with this idea, from a postmodern viewpoint, is that the so-called 'temporary' truths in Modern science have often reached such levels of acceptance and dependability--despite alternative suggestions and evidence explaining their inadequacies--that they have come to be very permanent (so much so that they serve as the basis of the construction of the narratives of many other fields of human endeavour, including those in less well-formulated younger sciences).[48] And when science speaks of its 'temporary' truths it does so with the claim that its 'temporary' truths are more truthful than non-scientific truths. Postmodernism would seek to arrest the arrogance of science and wrench from it its claim to be the only field of human endeavour capable of supplying us with useful models.

If other endeavours are allowed a claim to be able to model the universe (and all things in it), then the voices of difference and dissent are allowed to reign and all models will be encouraged to be viewed for what they are, mere models.

Another way to puncture independent truths yet still retain some form of political responsiveness based on knowledge and wisdom, is to state that for every situation there are multiple truths;. i.e.: there is more than one adequate/accurate representation of reality which can be acted upon as a physical, social or historical base. To avoid the vortex of absolute relativism we can still admit that while there is more than one truth not all of these provide an equal base upon which to enact the particular social and physical change that we desire (although telling the difference between the efficacy of any two supposedly unequal bases would, of course, be a difficult affair given the incommensurability of the many unshared opinions/assumptions of those bases).

In a world suspicious of absolute truth, metaphor plays an important role. Truths can be seen as metaphors that have just been sufficiently indoctrinated into language and culture that they have lost their status as metaphors. This Nietzschean view that 'truths are metaphors' would opine that we have long forgotten the parentage of our truths (from what sites the metaphors were originally casting comparisons from) and we have long forgotten the original purpose of their use.

We might even choose to go further than this--and put trust in the previous chapters--to decide that the 'original' metaphors were themselves not based upon grounded referents but upon other socially ingrained metaphors. Each truth that is spilled out into the universe from our mouths and our pens, therefore, is a metaphor based upon another metaphor.[49] There is no ultimate basis from which to make comparisons--and our truths ultimately float free and unattached above a never found, non-existent reality. What this means is that when we examine large-scale truths (like the unity worldview) as well as some smaller-scale truths (like the self-organisation of plant communities), then truths may be classified as attempts to describe the world by the use of metaphorical narratives that themselves are derived from elaborate, and now dead, metaphors of metaphors.

Some students of metaphor might charge that we do not have to stop at assuming that only large narrative structures like the 'unity of nature' or 'self-organization' are free-floating metaphors. Lakoff and Johnson for example, maintain that:

> our ordinary conceptual system, in terms of which we both think and act, is fundamentally metaphorical in nature. If...our conceptual system is largely metaphorical, then the way we think, what we experience, and what we do every day is very much a matter of metaphor.[50]

If we agree to dispose of the idea that there is, in either large- or small-scale metaphors, a grounded, reality-checked 'original' metaphor, we soon come to believe that every utterance we make, every conceptualisation we hold, and every action we take is dependent on an endless cascade of metaphors of metaphors of metaphors. A thing is elaborated not in literal terms as to exactly what it is but in terms of how it is like something else. And that something else is elaborated in terms of how it is like another something else (and sometimes in terms of how it is like the first something else). If large-scale metaphors such as the unity of nature exist in such a cascade then there should be no wonder that we find it impossible to locate the single origin of the unity metaphor and its supporting concepts.

What this emphasis on metaphors suggests is that the popular duality between 'the literal' and 'the metaphorical' is no longer tenable. It seems that rearranging and resurrecting metaphors (and hardening them into 'truths') gives us historical change and intellectual contingency. Rearranging and resurrecting metaphors also gives us politics since the rearrangement and resurrection of specific metaphors for political gain--rather than the search for the truth--is all that there seems to be.

Relating all of this to the Foucauldian Power/Knowledge thesis we might state that in Modernist culture the belief in Truth:

> is very much alive and truth is always absolute truth, the people who get to impose their metaphors on the culture get to define what we consider to be true--absolutely and objectively true.[51]

Rearranging and resurrecting metaphors is also the only activity of the human sciences, suggests Don Miller, since the human sciences:

> can only be engaged knowingly or otherwise in one enterprise: the creation of metaphors to analyse other metaphors.[52]

So how does all this relate to Postmodern Science? The main thing to note is that it seems that a lot of Postmodern Scientists are convinced that their worldview is not just a metaphorical rendering of the universe but a description of actual reality. For instance, in *On Purpose* Charles Birch indicates that although the Newtonian mechanist worldview is metaphor, Postmodern Science exposes an organic reality.

New Scientists, like Paul Davies, may be even more committed to the truth-rendering properties of the sciences of Chaos, complexity and self-organisation. For instance, of his own work Davies says:

> while emphasising the short-comings of a purely reductionist view of the nature, I intended that the gaps left by the inadequacies of reductionist thinking should be filled by additional scientific theories that concern the collective organisational

properties of complex systems and not by appeal to mystical or transcendental principles.[53]

Here Davies can be noted as explicitly defending both reductionism and the New Sciences as purveyors of scientifically-refereed reality while also extolling his commitment to the truth-exposing nature of the New Sciences compared to the inferior story-telling nature of traditional religious narrative.

New Scientists generally feel that their work is only subject to revision in the same way that all of science is subject to revision (in the vein of the Royal Society motto). This would tend to indicate that on the subject of truth and reality, Postmodern Science is not strongly allied to the rest of postmodern thought.

Absence (eg: Hassan, Bauman, Rosenau)[54]

For postmodern intellectuals of various disciplines a familiar theme within postmodernism is the rejection of the notion of 'presence'. In postmodern literary studies and art history this involves the rejection of the author or the artist as an authoritative meaning-generating presence in the works that they produce. It also over-turns the premise that art and literature must reflect some ultimate or absolute reality since the presence of such reality is taken away.

In academic studies about how humans come to know things, the predilection towards absence manifests itself in the rejection of an independent reality against which words and sentences can be measured and quantified.

In sociology and political studies absence has come to mean the absence of a fundamental social or political reality. Zygmunt Bauman explains sociological postmodern absence like this:

> 'postmodernity' being a semantically negative notion, defined entirely by absences--by the disappearance of something which was there before--the evanescence of synchronic and diachronic order, as well as of the directionality of change.[55]

One consequence of this acknowledgment of absence, according to Rosenau, is the postmodern predilection towards "absence of unity"[56] which believes that since all artistic, intellectual and social experiences and episodes are hardly united or unifiable, then no single unified theory can explain away all experiences and episodes.

So how does this postmodern notion of absence relate to Postmodern Science? It seems only negatively, since Postmodern Science is devoted to understanding nature in unity, a unity whose presence is taken to be both real and of deep consequence for the future of humanity and the Earth.

Aporia (e.g. Derrida and Bauman)[57]

Related to absence, and specifically the absence of unity that Rosenau talks about above, is the postmodern concept of aporia. Aporia is a word derived from Greek that translates into English as 'the absence of a passage'. It is a word that indicates the attenuation and loss of connection between things. Such loss must inevitably frustrate any attempt to unify things together.

In this regard the holistic unity of Postmodern Science is an inadequate reflection of either a) the absence of connections between things or b) the absence of unity all together. Whereas Postmodern Scientists like Charles Birch would defend the reality of the total interdependence and interconnection of all things in nature[58], and thereby attach supreme importance to the notion of the causal web--where one thing always affects another then another and so on--postmodern aporia would, at the very least, describe the existence of non-dependencies and non-connections in the causal web. It may even do more, going on, perhaps, to declare the death of the causal web through aporia. Postmodern aporia would then describe the world about which Birch's Postmodern Science narrates as being devoid of unity and connectivity. On the count of aporia, then, it seems that Postmodern Science fails to be postmodern.

Anti-communication (e.g. Barthes, Derrida, Lyotard, Kroker and Cook)[59]

It has been highlighted by many intellectuals how postmodernism involves an intense investigation of communication and the impacts of changing schemes of communication on culture and society and identity.[60] However, postmodernism is not necessarily a celebration of the ultimate efficacy of communication and indeed might be considered to offer a view of communication that declares that communication is, actually, a utopian intellectual promise that is never fulfilled. When a communication is made (by any one to any one else or by any thing to any thing else) it is never an exact communication; it is never a 100% sound and clear reproduction of the intended message. The process of communication faces many intrinsic barriers that get in the way of message fidelity. For example, language itself is metaphorical and suggestive rather than representative and literal, whilst the private linguistic and social experiences of each individual to any referent dissolves the possibility that two communicants are necessarily talking about the same things (as does the individual experiences of the communicants to a specific or purported referent).[61]

The power structure of the communicants also muddles messages as the communication thence becomes political. The schizoid and polysemous nature of most words must also be hurdled since, as Sim points out by paraphrasing Derrida: "any given word has a cluster of associations around it that undermine its supposed purity."[62]

We must also acknowledge the ideological components of many communication and language systems whereby a mode of communication or language is entrenched with a certain way of looking at things (as explored by Karshevskij[63]). Communication thus has to be undertaken within a changing medium by changeable communicants in schizophrenic and unequal social circumstances. Although one might gain satisfaction from a communication it is just as likely that a communication will leave one unsatisfied. The reality of communication is one of "radical isolation."[64]

So is this anti-communication sentiment reflected in Postmodern Science? To start with it seems fair to say that a Postmodern Scientist may able to operate within his or her worldview whether or not they adopt this postmodern stance of anti-communication. For instance, acknowledging that Power structures often make communication a barrier to, rather than a conduit of, the expression of certain lived experiences is possible whether you are an advocate of Postmodern Science or not. A proclivity to acknowledging the pervasivity of communication as a barrier (rather than a conduit) is not extinguished by buying into ecological holism, unitarianism and organicism. This would suggest that as far as anti-communication is concerned a judgment as to Postmodern Science's reputed postmodernism cannot be made.

However, if Postmodern Science is the intellectual child of systems theory (and it probably is) then there may be a link between communication and Postmodern Science that does entice the jury to reconvene and charge Postmodern Science as failing to be postmodern once again. Here, I wish to acknowledge that a primary theoretical position within systems theory is communications theory. As Lilienfeld says[65] "communications theory is largely concerned with the processes whereby messages can be coded, transmitted and decoded" in a whole variety of systems. Within communications theory, codes are transcribable in a literal way. What a biological molecule says when it is encoded in one cell is the same as what it says when it is decoded in another cell; and what a radio transmitter transmits in decoded form is the same as what a radio receiver receives. If Postmodern Science is keen to adopt this theory from their intellectual parents, then it too may find itself with a fetish for communication; marveling about the way things are communicated in both natural and artificial systems. Such a fetish may de-legitimise Postmodern Science's claim to be postmodern.

Interest in communication theory is, however, not a general characteristic of Postmodern Scientists. Where some New Scientists like Merry and Ayres are very fond of it, Postmodern Scientists like Birch and Gare are either unaware of it, or feel it has no place in their elaboration of a Postmodern worldview.

Pastiche (e.g.: Jencks, Harvey, Rosenau, Appignanesi and Lewis).[66]

In the art world the term pastiche refers to works of art composed of varied parts brought together, often haphazardly and arbitrarily. In postmodern art the style is developed to mean the conglomeration of all sorts of unrelated (and related) objects and styles, none of which acts as the focus or control or base upon which the others rest. This emphasis of defocus, un-controlledness, and baselessness means that no part of the art work can be called the focus or the essence of the work. Similarly when referred to in the world of the social, postmodern pastiche conveys a sense of the intermingled curdling of social space and social time. In any one place all sorts of cultures may pervade and then articulate upon others, be infused and defused by both similar and dissimilar politics, and project and reject all sorts of value frameworks. A social space is thus a mosaic of sometimes interlocking and sometimes independent ideas, practices and values. No one social theory can explain how all these parts work or how they came together. Nor can social change predicated upon these theories change the whole of society.

Again this characteristic of scholarly postmodernism may seem to only relate to Postmodern Science in the sense that they are mutually incompatible. Postmodern Science, as advocated by Ferre, Birch, Gare, Bohm, Griffin and others, seeks to work out how all the parts come together and totally interact to form a coherent whole. Clements' superorganism and Lovelock's Gaia are hardly pastiches but more like complete all-seeing murals of the natural world.

Chaos theory might be thought to be compatible with pastiche since pastiche is sometimes arbitrary and impervious to theorization. However, the arbitrariness of the chaos in Chaos theory is directly related to an impending, implicit or underlying order. This order takes away much of Chaos theory's claim to being a science of the arbitrary. It also imposes precise theoretical principles (negative feedback, order from chaos, self-similarity, fractal patterns, etc) upon its subjects of study, unlike the artistic and social pastiches envisaged by postmodern art and postmodern sociology. On the count of pastiche, then, Postmodern Science fails to be postmodern.

Fragmentation (e.g.: Heller and Feher; Harvey, Ashley and Lewis).[67]

When he was describing his Postmodern worldview, the Postmodern Scientist, David Bohm, stated:

> We have seen that fragmentary thinking is giving rise to a reality that is constantly breaking up into disorderly, disharmonious, and destructive partial activities. Therefore, seriously exploring a mode of thinking that starts from the most encompassing possible whole and goes down to the parts (sub-wholes) in a way appropriate to the actual nature of things seems reasonable. This approach tends to bring about a different reality, one that is orderly, harmonious and creative. For this actually to happen, however, a thorough-going end to fragmentation is necessary.[68]

How do we reconcile this with David Harvey's rendition of a postmodern worldview?:

> I begin with what appears to be the most startling fact about postmodernism: its total acceptance of the ephemerality, fragmentation, discontinuity, and the chaotic that formed the one half of Baudelaire's conception of Modernity. But postmodernism responds to the fact of that in a very particular way. It does not try to transcend it, counteract it, or even define the 'eternal and immutable' elements that might lie within it. Postmodernism swims, even wallows, in the fragmentary and the chaotic currents of change as if that is all there is.[69]

Harvey does not deny that Modernity and Modernism can be--and is--conceived as a period and a mindset where change and fragmentation is a persistent phenomenon. He does however feel that within postmodernity and within the postmodern mind such fragmentation and change is hardly intermixed, underpinned by, or contributory to either integration or stability. In the Modern world, change and fragmentation is ever-present but so are integration and stability. Not only that but Modernism has always utilised fragmentation and change as factors to contribute to ongoing changeless essences such as Man, Freedom, and the endless Progress of science, technology or the economy. Change is thus linked to technological and economic development, two icons of the Modernist age. Fragmentation within this social and cultural framework is usually conceived as part of modernisation and improvement. Within such a schema, the popular acknowledgment of the fragmentation of community and society into individuals is very important since such fragmentation is regarded as an important key to the growth of individualistic culture and entrepreneurship, and thus capitalist development.

If fragmentation and change are recognised as characteristics of Modernity, they are always linked to adjunct Modernist concepts like Progress, Mankind and Freedom. All these concepts were born in Modernity and died--without changing much--by the time of Modernity's supposed end; i.e. around about now.

When reconciling Bohm and Harvey we might note that Bohm was talking of the natural world and humanity's relation to it and that Harvey is talking of the social world and the human project to explain it. Harvey and Bohm are naturally more interested in the social and natural worlds respectively since Harvey is a social scientist and Bohm was a natural scientist and if engaged in conversation about fragmentation they probably would have talked past each other. Where Harvey would talk of the fragmentation of Modernity in terms of social fragmentation and of postmodernity in terms of cultural, social and epistemological fragmentation, Bohm would have preferred a simple binary division: Modernity equals fragmentation and Postmodernity equals an end to fragmentation.

For many postmodernist scholars this binary division of Bohm's (Modernity is that period where fragmentation exists in our views of nature and Postmodernity is that where nature's unity is recognised) is either interpretable as 1) two strands of Modernism--industrialism and romanticism--fighting it out, with romanticism the rightful victor, or 2) as a clear cut case of Modernism and anti-Modernism. In this regard we could abruptly state that Bohm has just got his postmodernism wrong. His repulsion towards fragmentation is hardly a postmodern characteristic. This observation compels me to declare that on the count of fragmentation, Postmodern Science, once again, fails to be postmodern.

Anti-representational (eg: Lyotard and White[70])

> I only want to speak out for Gaia because there are so few people who do, compared with the multitudes who speak for people.

So says James Lovelock in his second book on the theory; *The Ages of Gaia.*[71] Through his geophysiological science of the biosphere Lovelock claims to have unearthed the real nature of life and to offer an appropriate avenue whereby the actual needs of the whole Earth, which were for a long time hidden from environmentalists, are now displayed with abundant clarity.

As we can see from the statement above, Lovelock claims to represent the Earth. Herein lies another problem explored by postmodernists, that of representation. Representation, it is said by postmodernists, is never feasible because representation always involves 're-presentation'. Whether we are talking about the epistemic representation of the Truth by science, the political representation of the people by elected politicians, or the cultural representation of identity through national and social mythologies, representation is seen by postmodernists as an insidious cover for non-participation, elitism and *status quo* politics. Given this, the Earth's laws, desires and needs are not so much represented in Lovelock's Gaia theory so much as they are 're-presented' via particular intellectual, political and philosophical biases that affect Lovelock and his theory.

Whereas Modernist social progress has been advocated time and again as a widening of representation in various social settings, "postmodernists believe that representation encourages generalisation."[72] Indeed, this is what we find in Lovelock's representation of Gaia since the nuances and plurality of the world's ecological communities are overshadowed by preoccupations with the big picture. In this process of generalisation, Lovelock's claim to represent life in Earth are helped by the abstraction of the category of 'Life'. Just as categories and abstractions such as Reason, Man and Truth are too general to account for specific reasons, humans and truths, respectively, so Lovelock's use of Life is over-generalising. In this way when Lovelock talks of Gaia maintaining the planet for Life he is re-presenting the lives of species with the Life of Gaia. For many a postmodernist this victory of the abstract over the concrete is a hallmark of Modernism. If Gaia theory is inherently part of Postmodern Science, as claims Merchant, Ferre and Birch, then Postmodern Science fails to be postmodern.

Otherness (eg: White, Harvey and Tester[73]).

David Harvey declares that:

> the most liberative and therefore the most appealing aspect of postmodern thought [is] its concern for Otherness.[74]

Otherness is usually described by postmodern sensibilities as that category of objects and subjects which do not conform with, contribute to, or admire the mainstream majority, mainstream Reason or mainstream knowledge. Usually this refers to those members of humanity whose histories, values, traditions and lifestyles have been described (by that mainstream) as abnormal, strange, savage, uncivilised or unnatural, and whose voices have been marginalized, patronised, forgotten or have been filtered through mainstreamism. As far as the living world is concerned all non-humans in the Modernist world have been treated in this way and so they have been sub-consciously branded by Modernist sensibilities as the 'Other'.

There is often a case made that 'Nature' itself has been badly treated by humans in this way and, as a whole, is confined by humanity to the category of being an

'Other'. Thus it can be claimed that if you are protecting, speaking for, and celebrating Nature in any way, shape or form you are celebrating 'Otherness'. Thus Lovelock can, and does, claim that he is speaking for the silenced Earth as though it is a marginalised, forgotten entity.

However, Lovelock's role as spokesperson for the planet Earth can also be interpreted as shunning Otherness. Instead of bringing every living thing into the fold of the united global Gaian system, a postmodern approach might look at the differences contained within the system; identifying how the parts are different from each other rather than the same. When identifying the uniqueness of the parts, their Otherness is in some way released. When generalising the activities and properties of the parts under the rubric of concepts such as the 'superorganism', the 'ecosystem' or the 'global system' Otherness is washed away. If Postmodern Science allies itself to the celebration of these unitary concepts then it, too, can hardly claim to be an effective celebrant of Otherness. On the count of Otherness, then, Postmodern Science fails to be postmodern.

Intertextuality (eg: Barthes, Riffaterre, Rosenau[75])

As indicated by Charles Birch in *On Purpose* one of the 'doctrines' of a worldview derived from Postmodern Science is interconnectionism. It may be pointed out that there is a potential articulation between Postmodern Science and postmodernism through their respective focus upon interactionism and intertextuality.

Intertextuality, in postmodernist circles, is the acknowledgement of the interconnected processes that go on during the making up of a text:

> Writers who create texts or use words do so on the basis of all other texts and words that they have encountered, while readers deal with them in the same way.[76]

This means, as Stephen Heath makes clear, that a written work is not a self contained, individually-authored whole but the absorption and transformation of other texts. Heath goes on to cite Julia Kristeva to edify this idea: a text is "a mosaic of quotations."[77]

Now this would only seem to relate to Postmodern Science in a tangential way since the interactionism supposed in postmodern intertextuality is only pertinent to written documents. However, under postmodern analysis all things can be read as texts, whether written/printed or not. Therefore everything that humans encounter in their daily lives; television, radio, food, drink, pedestrian crossings, the sky above; they all become texts, and this applies equally to those things we encounter within nature. In this way, nature, and the things in it, can exist in:

> infinitely complex interwoven relationships. Absolute intertextuality assumes that everything is related to every thing else.[78]

Postmodern Science can thus possess the postmodern attribute of intertextuality because of its penchant for interconnectionism. We may also, at this point, note Rosenau's statement that:

> Everything comes to be defined as a text in a postmodern context, and yet the text is marked by an absence of any concrete or tangible content.[79]

This statement can be compared with Charles Birch's idea that there are no objects and with the general Postmodern Science idea which emphasizes events and processes as primary compared to substances and objects.[80] From such a comparison we could imagine some common theoretical ground between postmodernism and Postmodern Science. However, whereas Birch would maintain that these ideas refer to real things out there in the real world, postmodern intertextualists would be inclined to believe that their texts and their intertextualities are merely maps of an impossible-to-know, or non-existent, reality. Texts such as Birch's five doctrines of a Postmodern Worldview would be classified by 'true' postmodernists as socially constructed human attempts to classify, describe and change our maps of the world rather than actual reflections of the real world. Thus rather than believing that atoms on Jupiter are interconnected with gorillas in the misty mountains of Rwanda--as Birch suggests in his doctrine about the total interconnectedness of all things–postmodernists would say that gorilla's are socially-constructed beings whose image in human minds may be dependent upon the atomic theories, scientific metaphors, cultural processes, and intellectual contingencies that describe the physics of Jupiter.

We may also be able to delineate another possible difference between Postmodern Science's holist interactionism and postmodern conceptions of intertextuality. Whereas Postmodern Science thinks of things as being so overwhelmingly interconnected that they constitute a unity, postmodern intertextuality allows for holes in the net of intertextuality; areas where there may be isolation between any two texts. If Postmodern Scientists were to adopt similar ideas for their Postmodern Science they would have to acknowledge the incompleteness of the world's physical and ecological interactions.

One aspect of interactionism that has been utilized by Postmodern Scientists is the concept of 'the unity of the observer and the observed', as epitomised in Bohm's view of quantum physics. The familiar wave-particle thesis was thought, by David Bohm, to be antithetical to Modernist sensibilities because the thing-ness of a thing is determined by the context of the observer. Observed one way, an elementary particle does, indeed, appear to be a particle, but observed another way, it appears to be a wave: "the quality of the thing..." said Bohm "...depends on the context."[81] (For Wiessert this observer-dependency--and the unity between observer and observed that it suggests--tells us that quantum theory, despite arising within the Modernism of the early Twentieth century, is actually a part of Postmodern Science.)[82]

So how does all this relate to postmodernism? Well, the observer-dependency of quantum physics might strike some postmodern readers as being comparable to the postmodern predilection towards the elimination of a context-independent reality as outlined in the above subsection about *Anti-truth and Anti-realism.* Would this, then, convey a certain degree of complimentarity between the postmodern intellectual movement and Postmodern Science? Probably not, since the context-dependency of quantum physics as described by Bohm and many others is a purely physical phenomenon dependent upon the physics of the equipment used to measure the subject of its studies (i.e. fundamental particles); whereas critical postmodern views of context-dependency are thoroughly social and cultural in nature. The thing-ness of a thing in 'postmodern' quantum physics is dependent on the equipment used to measure it but is otherwise invariable whereas a 'proper' postmodern analysis of context-dependency would bring in those micro- and macro socio-cultural and intellectual factors that give rise to varying interpretations of reality despite the use of exactly the same equipment.

Chaos (e.g.: Harvey, Hassan, Grant, Lewis).[83]

Another one word characteristic of postmodernism might be 'chaos'. Postmodernism finds that differences in social situations and in the natural world explicitly reflect the implicit chaotic nature of the social and natural universe. If this is so, then it might be thought that Chaos theory is immediately relevant to postmodern theorising. Indeed, this is the declared view of many Postmodern Scientists:

> Advocates of postmodern science claim that the modern scientific paradigm...is giving way to a new paradigm based on principles of indeterminacy, chaos and evolution.[84]

Wiessert is an example of one of these 'Chaos theory is postmodern' advocates and he talks about the 'postmodern revolutions in Twentieth century science' presented by Chaos theory.[85]

As we have seen in the previous chapter, Chaos theory is held to be the scientific manifestation of the order from chaos idea:

> new forms of order emerge out of chaos, albeit considerably more complex, intricate and irregular in nature than the forms of order which previously were thought to exist.[86]

Chaos theorists have often derided 'society's' fixation with the goodness of order and the badness of chaos. This is a mistake, they believe, because chaos is a friend to the world since it gives rise to order. Chaos and order, Chaos theorists say, are connected and complimentary: chance is a part of purpose, stability is achieved through change, and order is hidden within chaos. As Best explains it, the chaos that Chaoticians are fond of is not perceived as the logical antithesis of order, for that is the type of rigid binaristic thinking that postmodern science attempts to displace.[87]

Although Chaos theory might disrupt the binary nature of the chaos/order dichotomy in Modernist thought, it does this in a thoroughly Modernist way; by having order take over and dominate chaos. Although Chaos theory has been claimed as a Postmodern Science and as a celebrant of chaos, the chaos in Chaos theory is a bounded and packaged user-friendly chaos, a chaos that can be managed, controlled, utilised and functionalised. The chaos of Chaos theory is dished out in small doses--like vitamin pills--to help develop orderly, progressive processes.

Chaos theory is Modernist, not postmodern. Modernists try to limit disorder and contingency; and this is what Chaos theory, despite its claims, is largely about. 'Proper' postmodern chaos, if it were theorised, would not be bounded so. It would flourish uncontrollably. It would be a chaos constituted by the happenings of local agents whose collective pattern is sometimes interesting but hardly meaningful as a base for foundational views or political action. Postmodern chaos cannot be packaged as some sort of subsystem of variability within a mainframe system--and it exists precisely because it cannot be systematised and packaged.

This need not be a reinstatement of the binary nature of order and chaos as Stephen Best might allude, it is an acknowledgment of the pluralisation of both sides of the 'chaos versus order' dichotomy. Where traditional conceptions (i.e. Modernist conceptions) of chaos and order have presented them as a *duality* (i.e. have set them up as being opposites), and where Chaos theory presents the chaos-order relationship as a *union* (where chaos contributes to creating order), deconstructive postmodernism might see order and chaos, and the order/chaos duality, in *plurality* so that there are many types of order and many types of chaos.

The chances are, however, that postmodernists might not even be prepared to pluralise order and chaos. Instead they might wish to get rid of the concepts altogether, acknowledging them as no more than power investing and power divesting categories.

While it is possible to respect the argument made by Chaoticians that under Modernism, chaos is regarded generally as a bad thing, the problem still remains that their particular chaos is such a pathetic chaos that they, too, appear to want it restrained by order. In this way it seems as though Chaos theorists covertly regard chaos as bad too. With a name like 'Chaos theory' and with its avowed interest in chaos you would think that Chaoticians were setting up a celebration of all those things that have come to denote chaos; including for instance; anarchy, uncontrolledness, wildness, etc. This is hardly the case, however. As intimated before Chaos theory is more of a continuation of the Modernist celebration of order and how it is achieved than an advocate of its supersession. If Chaos theory only venerates and values chaos because it contributes to the natural order of things, then obviously chaos is not the only thing being valued; both order and chaos are valued. But they are valued in different ways. Order is valued intrinsically, and chaos is valued instrumentally. Thus order of any type is valued in and of itself; a manifestation of the universe's essential goodness, whereas chaos is valued because it is instrumental in the making of order.

We can detect this varying valuation of order and chaos in the following statement by a Chaotician: "Deep chaos is a natural, inescapable, essential stage in the transformation of all life."[88] Here chaos is presented as a phenomena that should be valued not for itself but for its contribution to all life.

What this means is that chaos, and those things deemed to be chaotic, appear to exist as secondary entities to those things deemed to be orderly. In the ecological arena this means that those things which are held to be chaotic--which for many 'order from chaos' theorists often means such things as the behaviours of individual organisms and the groups they form--are only of value because they contribute to orderly ecosystems and then the overall Gaian order.

To be classed as postmodern, any competent postmodern scholar must surely give full parity of value between chaotic and orderly things; chaos must be regarded as a valuable whatever its relationship to order. Postmodern Science does not do this.

Another way that Chaos theory fails to live up to the claims by Postmodern Scientists that it is a postmodernisation of science has been made by Rosenau. Despite being hailed as radical, revolutionary and anti-Modern science, Chaos theory nevertheless originated within traditional mathematical and physical sciences. Its formalistic nature, and its presumptions towards the practical problems of Modern physics, thus enfolds it into the traditional techno-scientific framework. Or, as Rosenau points out:

> chaos means not the end of modern science but rather a 'new science' that improves on the old and makes it even more powerful.[89]

One particular proponent of the 'Chaos theory equals Postmodern Science' equation admits as much, stating:

> Chaos theory is not opposed to normal science. It is normal science. Its criteria for evaluating evidence, reproducing results, credentialling investigators, and so on, differs not at all from these other physical sciences.[90]

Not only this but Chaos theory and its relatives in the New Sciences such as complexity theory and self-organisation theory are delivering messages of Baconian utilitarianism in the same old way that Modernist science did:

> Complexity is on the cutting edge of science. Physicist, Heinz Pagels put it this way: 'I am convinced that the societies that muster the new sciences of complexity and can convert that knowledge into new products and forms of social organisation will become the cultural, economic and military superpowers of the next century.'[91]

Incoherence (e.g.: Bauman, Babich)[92]

According to Babich postmodernism can be seen as "running up against the modern ideal of clarity and distinctness."[93] This, for some, is interpreted as a negative thing and for others it is a positive.

Coherence serves as both a tool for action but also as a prison. It enables policy formation since it offers views upon which action may be taken but it also codifies into a unitary form the widely divergent phenomena and experiences within the world; rejecting difference and atypicality so that something can be said about some situation and action can be put into effect.

When examining the relevance of coherence to unitarianism we see that Edward Goldsmith, for one, tells us that coherence is a most definite plan of his 'post-industrial worldview':

> I have tried in this book to state clearly the basic principles underlying an ecological worldview. These principles are closely interrelated, forming an all-embracing and self-consistent model of our relationship with the world in which we live.[94]

Likewise, Paul Davies seeks to realise the inherent coherence of all of nature's operations when he talks of the laws of the nature 'dovetailing' neatly together to give rise to a coherent whole.[95]

Neither Davies, nor Goldsmith use the term postmodernism in their elaboration of the worldview of the New Sciences but their views on the meaning and operation of the universe are supported by many of those that do (for example Birch, Gare,

Griffin, Ferre). These Postmodern Scientists also implicitly testify to the importance of arriving at a coherent philosophy of nature.

What all this means in this analysis is the unearthing of yet another divergence between Postmodern Science and postmodernism. Whereas Postmodern Science wants to codify and integrate a precise worldview upon which ethical changes can soundly rest, postmodernism might choose to explicate wholly variable, multiple and sometimes contradictory worldviews in which the objects and subjects therein contained are also thought of as being variable, multiple and self-contradictory. In this light, a single view of nature as unitary, holistic, interconnected (and a single view of all the various processes of nature as self-organising orderings) may be coherent and understandable concepts but they are also imprisoning ideas since by imploring us to believe that the universe--and all life in it--is a single self-organising, holistic system, other ways that the universe may be understood and interpreted are de-legitimised.

Anti-teleological (e.g.: Slack & Whitt)[96]

After observing the order in Gaia and her ecosystems and after observing the evolutionary legacy of the members of Gaia and her ecosystems, Paul Davies declares that life on Earth "has a pleasing teleological character to it."[97] Charles Birch, a self-declared Postmodern Scientist reiterates the exact same point over and over again in his 1990 book.[98] This is a book designed to counter what Birch perceives as a predilection towards meaninglessness in science--and the meaningless human lives that this type of science supposedly leads to. Reconciling Davies' and Birch's focus on teleology within their Postmodern Science with more general postmodern sensibilities in academic writings is not easy since many postmodern writers would be classified as being avowedly anti-teleological in their approach. Such postmodern suspicion of teleology arises from a variety of quarters; the association of teleology with Progress being a primary one.

Postmodernists hold that Progress has become such a mythic emblem within Modernism that it is hard to contemplate social change without the word and the idea being raised. Technological and economic progress are particularly relevant here. Grenz for instance, declares that postmodernists are involved in "eschewing the enlightenment myth of inevitable progress."[99] Progress, the idea, is thus described by postmodernists as a product of Modernism and it manifests itself in glorious fashion when pertaining to socio-economic, technological and scientific advancement. To abandon Progress (as either an essential or desirable element in human society) as the postmodernists have:

> is to accept the idea that history may have no purpose, that it is not an evolutionary or progressive march towards an emanicipatory telos, but rather a contingent set of events, often accidental and with many unanticipated consequences.[100]

When dismissing Progress, postmodernists are usually talking from within disciplines dealing with human agents and processes, for example history, philosophy, development studies, sociology, cultural studies and the history of science. Within all of these disciplines, Progress has been abandoned by the resident postmodernists and along with it any teleological arguments which state that what has gone on before (say in society) is leading in an identifiable direction (say towards more social complexity, higher living standards, greater knowledge, expanded freedom and so on).

Science itself has long been hooked up on the idea of Progress since it is useful as a rhetorical and ideological tool when extolling the virtues of the discipline: 'science is forever expanding the domain of human knowledge and feeding this ever-increasing knowledge into practical benefits for the betterment of all', or so scientists generally say. Postmodernist pessimism, as outlined earlier, doubts that science does this at all and they suggest that the knowledge science does come up with is just a politically-obliging or intellectually-convenient set of contemporary stories and metaphors that enable scientists to convince others whatever they are doing is socially and economically useful.

When it comes to the subjects of study that it looks into, science is also hooked up on the Progress idea. In biology, for example, this progress is encapsulated in various evolution theories, most notably 'evolution by natural selection'. Biology is not alone, of course; cosmology, geology, physics, chemistry etc all have their own affiliations with evolutionary ideas. Within many of the disciplines of these sciences there lies a tendency to be amazed about how complex things evolve from simpler things.

While many scientists try to eject teleology as an explanation within evolution, most Postmodern Scientists are not so adverse to the use of teleology and purpose when describing the activities of nature. Birch, Davies and Goldsmith, for example, feel that it is an integral part of their scientific and metaphysical descriptions of the world. If this is so, then there is obviously a sharp difference of opinion between Postmodern Science and postmodernism, generally. On the count of anti-teleology, then, Postmodern Science fails to be postmodern.

Heterogeneity (e.g. Hassan, Heller and Feher, Zagorin, Sim)[101]

Postmodernism seeks to reveal reality as a multifarious assemblage of varying elements. It diversifies and multiplies the oppressions and concerns of people and the angles of attack against such oppressions and concerns. Postmodernism recognises that single master models of social being and social development must give way to, what Foucault would say are, heterotopian images of what is and what can be. In a world acknowledging heterogeneity we must, says Foucault "prefer what is positive and multiple, difference over uniformity, flows over unities, mobile arrangements over systems[102] In heterotopia there can be little room for confining belief in just a single idea or a single God, and there is also little sympathy for an all-embracing metaphysics and for universal laws and principles.

So does Postmodern Science allow for such extravagant heterogeneity and pluralism? Certainly Postmodern Science's allegiance with unities would not convince Foucault, Bauman or Lyotard since these postmodernists have taken as a prime problematic the need to disintegrate the systems concepts and their homogenising tendencies. Charles Birch's dismissal of the idea that the universe can be considered fractured, and his entrenched commitment to a single universe and a single way of looking at it, would, in this regard, seem to place him firmly within the Modernist mode. And along with Birch are carried many other Postmodern Scientists who would, likewise, tend toward very similar single and unitary views of reality.

Given this different attitude with regard to heterogeneity, it seems likely that Postmodern Science fails, once again, to be convincingly associated with postmodernism.

Anti-hierarchical (e.g.: Hassan, Deleuze and Guattari, Bauman)[103]

Bauman believes "postmodernity...is about flattening of hierarchies, absence of discretion and equivalence."[104] This again imposes a distinction between Postmodern Science and postmodernism because Postmodern Science, as we have seen, involves a promotion of a self-organising united universe which, as we have also seen, is necessarily inscribed with hierarchy. Postmodern thought, instead of ascribing nature and society in hierarchical terms, would attach itself to non-hierarchical conceptualisations such as Deleuze and Guattari's 'Thousand Plateaus'[105] (where knowledge and the things it refers to are disseminated into multiple planes, none of which are more basic or of higher stature than others). On the count of being 'anti-hierarchical' also, Postmodern Science fails to be postmodern.

Meaninglessness (e.g. Hassan, Harvey, Bauman)[106]

If we allow only one thing to characterise Postmodern Science then that one thing might well be its search for ultimate meanings. If we allow only one thing to characterise postmodernism then that one thing might well be its rejection of ultimate meanings. For postmodernists, ultimate meanings, and the search for them, totalise the myriad realities of the world into restrictive schemes that can hardly do justice to the differences of opinions, ways of living and changing experiences that abound.

For Postmodern Scientists, the search for new ultimate meanings must go on in order to free us from the meaninglessness of the Modernist universe and to overturn oppressive Modernist ideals:

> if there is no formulation to the meaning of our life, we are adrift. One of the agonising dilemmas of our times is the death of meaning. The relentless march of the empiricist worldview has denuded us of meaning.[107]

Charles Birch is one of the Postmodernist Scientists who has been searching for the meanings that have been ejected by Modern science. Here, again, Birch's focus on meaning is unitary and singular; there is only one meaning and all must follow it. Birch admits that the universe is in some way composed of numerous individual entities that appear to be manufacturing their own experiences and their own meanings but about this he states:

> A multiplicity of creative agents implies the need for the rule of one. Too many cooks spoil the broth. There must be something that sets limits to the confusion and anarchy possible with a multiplicity of creative agents. Individual purposing agents need to be co-ordinated.[108]

These two sentences tend to express exactly what Postmodern Scientists are willing to give up in the search for their ultimate meaning, namely difference, dissent, and confusion. The need for a grand and ordering meaning takes away the individualistic nature of meaning. Meanings can only exist as large-scale foundations for Postmodern Scientists, not as small-scale pragmatic schemes of thought which are merely part of the changing mindscape in the lives of groups of individuals.

Both Henryk Skolimowski and Charles Birch try and characterise the lack of meaning as a Modernist project but Modernism is replete with many founding principles that are equivalent to meaning: Progress, Truth, Unity, Freedom, Mankind; all of these concepts are positioned within Modern science as fundamental and all are hailed as meaning-invested conceptualisations replete with a spiritual/religious overtones (to the point that people not affirming them are held out as heretical).

Another reason for requiring ultimate meanings to be discovered and disseminated, according to Postmodern Scientists, is to replace environmentally and socially detrimental ultimate meanings with new environmentally and socially benign ones. On this score we can note how many philosophers of nature (including most Postmodern Scientists) have attempted to wrestle the 'struggle for existence' metaphor away from nature and expose it as a bourgeois interpretation of the natural world which justifies the *laissez faire* capitalism of Modernism. In its place they seek to imbue nature with an intrinsic co-operative meaning; upon which they wish to realise their varying visionary models for a post-industrial or postmodern ecological society.[109]

A thoroughly postmodern appraisal of the situation, however would not attempt to describe nature with any meaning. Although a meaningless nature casts our metaphysical outlook on life adrift from any cosmic purpose, at least we are not enslaved into the metaphorical ravings of those who attempt to exercise control over us by appealing to the natural law of things. Each individual develops its own meaning of nature (though this is usually mediated by its socio-ecological context) whose precise characteristics are unique to that individual. The unique contextualised experiences of each individual and the absence of totally transcribable true representations and communications of those experiences (due to the inherent difference in each individual's contextualised experiences) means that no two individuals can have the same compliment of experience-generated meanings. Every individual thus carries around its own hermeneutical universe. Meaning is thus localised and de-centred into the mini-universes that exist as each individual. Where meanings do overlap from one individual to the next, there are plenty of meanings to differentiate them.

Something to note here is that such individuation of meaning within the human realm does not have to lead directly to nihilism and the abdication of social responsibility (something many a Modernist and postmodernist alike often allude to as being the case). While meaning is individualised, the inter-subjective abrasion of meanings--via the collision and conflict of clashing universes--is a prime medium for the voices of difference to be heard. An over-arching meaning acts to stifle this abrasion and silence the exchange that flows from inter-universe conflict.

Because "the postmodern perspective reveals the world as composed of an indefinite number of meaning-generating agencies, all relatively self-sustained and autonomous, all subject to their own respective logics and armed with their own facilities of truth validation"[110] then to prescribe meanings for other universes is an act of metaphysical imperialism; an attempt to colonise another universe and reconfigure its reality.

To construct a particular underlying or ultimate meaning for nature (whether or not that meaning is based upon unity) has a tendency to paint over the many instances in nature where unity, and any other ultimate meaning, is completely irrelevant. Such neglect would defocus empathy with the lived experiences of individual organisms and the collections they comprise. Somehow there must be a willingness to allow multiple sets of meanings and multiple sets of purposes in the universe rather than one ultimate meaning and one ultimate purpose.

For many postmodernists the idea of meaning is itself a dangerously stifling and totalitarian concept. It places singularity and closure on whatever it is referring to; a sentence, a social pattern, or the whole entire universe. For postmodernists the only way out of such singularity and narrowness is to get rid of the idea of meaning altogether, including the ultimate meanings that Postmodern Scientists believe are floating around in the universe waiting to be uncovered.

As an example of the tyranny of meaning, consider the writings of management specialist Margeret Wheatley. In her book on Chaos theory and organizational studies she asks the question "Is there a magnetic force, a basin for activity, so attractive that it pulls all behaviour toward it and creates coherence?" Yes, she answers, by saying:

> my current belief is that we do have such attractors at work in organisations and that one of the most potent shapers of behaviour in organisations, and in life is meaning.[111]

The way Wheatley equates attractors to meaning is a telling connection between Chaos theory and its closeness to unifying and universalising thought. Remember; Wheatley's theories are an attempt to break down the monolithic structure of organisations and their decision-making and action-taking systems. When examining organizational liberty in her book, Wheatley feels it important to bring in the concept of 'meaning'. Suppose we think of 'meaning' as a word given over to describe a 'unity of purpose', a 'reason for being'--not an unreasonable idea given contemporary popular conceptions of what 'meaning' is said to portray. This, in relation to Wheatley's statements above, suggests that she is claiming that the behaviour of parts of an organisation are governed by a 'unity of purpose' or 'reason for being'.

If this is so, then the important question to ask here becomes: 'where does this leave freedom and autonomy in these organisations?' Freedom and autonomy have been granted if, and only if, it conforms to the 'natural' boundaries of the strange attractor that Wheatley calls meaning. This might be all right if the organisation exists under a meaning that is imbued with sentiments that indicate the intrinsic value of those who work in the organisation and how they may be kept happy and contented. However, the 'meaning' of an organisation, especially in the early Twenty First century is almost always more likely to revolve around either profit or efficiency, than it is around worker fulfillment.

Under the operation of an ultimate meaning like profit, individuals who do not contribute to this meaning cease to be part of the organisational system and will be either made to come back into line or be expunged from the system altogether. What this means for those entities that make up the system, otherwise known as people, is that they have to do what the management says or be disposed of. If Wheatley's 'postmodern' organisations are to eschew this totalitarianism, and if organisations are to be made more able to deal with freedom and localism, they must eschew the existence of a primary overarching meaning and realise that an organisation is full of highly dispersed and highly contingent meanings, none of which should be necessarily applicable to all members.

From this analysis, which suggests Postmodern Scientist's may be far too fond of 'meaning', it is necessary to conclude that Postmodern Science fails to be postmodern.

Anti-foundational (e.g.: Hassan, Harvey, Best and Kellner, Lyotard)[112]

As Best and Kellner point out, postmodern theory finds that:

> Modern theory--ranging from the philosophical project of Descartes, through the Enlightenment, to the social theory of Comte, Marx, Weber and others--is criticized for its search for foundational knowledge, for its universalising and totalising claims, for its hubris to supply apodictic truth, and for its allegedly fallacious rationalism.[113]

Although it is often critical of rationalism, Postmodern Science nevertheless stands within Modern theory as Best and Kellner have fashioned it in the above quote.

If Enlightenment thinkers searched for principles that could be placed at the foundations of all social and philosophical theory, so Postmodern Scientists search for the foundations of a true description of the world which may forever-more be held as basic knowledge. As has been revealed above, the foundations for Postmodern Science lie within unity and it is upon this foundation that new scientific knowledge, ecological insight, environmental evaluation and social policy must be crafted. If the characters of these latter human endeavours may change, waxing and waning to and fro via historical contingency, Postmodern Scientists like Griffin, Birch, Gare, Ferre and Artigiani are nevertheless determined to embed unity as the bedrock foundation of all metaphysics.

If, as Haney alludes in his studies of postmodernity[114], anti-foundationalism is an openness to the questions of 'is-ness' of the world, then Postmodern Science must be interpreted as belonging to Modernity since it has closed down this avenue of debate by stating and restating that the 'is-ness' of the world is one of undeniable unity.

Another obvious problem for Postmodern Scientists revolves around the enunciation of alternatives. Postmodernism 'proper', it is usually suggested by its supporters, should aim to offer an explanation of current worldviews without actually providing an alternative worldview. To do so, say postmodernists, would be to harden a whole plethora of biases, metaphors and illusions into dangerous theoretical unities that might soon act as foundational statements; all casting down their own sermons and policy-plans. Alternative foundations (in the study of nature or of society) are just not an option for most postmodernists.

Postmodern Scientists, however, jump easily from critiquing certain foundations (that of atomism, for instance) into erecting another in its place (that of holism for instance). This active program for the construction of an alternative foundation--with which all perspectives of the universe must in some way associate themselves--shows that, on the count of anti-foundationalism, Postmodern Science may, yet again, fail to be postmodern.

Anti-system (Bauman, Lyotard, Foucault, Sim).[115]

> The first victim of advancing postmodernity was the invisibly present, tacitly assumed spectre of the system, the source and the guarantee of meaningfulness of the sociological project and in particular, of the orthodox consensus.[116]

Pertaining to Bauman's above comment, there seems to be a sharp contradiction between Postmodern Science and postmodernism. While postmodernists have sometimes expressed intellectual outrage against the unifying assumptions implicit in systems thinking, Postmodern Science is actually dependent on, descended from, and completely at ease with systems thinking. Whereas Postmodern Scientists such as Artigiani, Birch and Gare (and others) advocate one form or another of systems thinking, postmodernist Zygmunt Bauman, in contrast, states:

> postmodernity would do well if it disposed of concepts like the system in its orthodox and organismal sense...Suggestions of a sovereign totality logically prior to its parts, a totality bestowing meaning on its parts, a totality whose welfare or perpetuation all smaller (and by definition subordinate) units serve; in short a totality assumed to define....the meanings of individuals....that compose it.[117]

Postmodern Science as Postmodernism: The Verdict

What the above section tends to show, then, is that Postmodern Science is hardly postmodern. Instead of recognising heterogeneity, diversity, difference, or meaninglessness (all of which are acknowledged characteristics of postmodernism) and instead of rejecting foundationalism, systems and teleology (which are rejected by most academic postmodernists) Postmodern Science promotes systems, defends unity against fragmentation and aporia, and revels in the search for foundational meanings (including the ultimate meaning of life and the universe).

To be sure, you do not have to exhibit all of the above characteristics to be postmodern, and many postmodern studies have little to do with some of the various specific characteristics listed above but any strong claim towards postmodernism might be thought capable of generously embracing at least a few of the ideas in some way.

Postmodern Science could be seen to reflect aspects of some particular postmodern characteristics listed above, intertextuality perhaps, but surely if you can qualify as being postmodern by exhibiting a tenuous link with just one particular characteristic of postmodernism then might you not be disqualified from postmodern membership if you exhibit the antithesis of all the other of those characteristics?

Intellectual Imperialism/Scholarly Totalitarianism.

We have already seen how Gaia theory exhibits a tendency for spatial imperialism but we can also raise questions with regard to its intellectual imperialism. When Lovelock speaks of modern science he says:

> The Victorians were careless when they allowed science to divide and become an array of sectarian expertises. Each newly separated science soon developed its own argot and gang of professors who ruled, from the cloistered turrets of their universities, over sharply bounded fiefs.[118]

As we shall see, however, far from attending to a destruction of these cloistered fiefdoms of knowledge, it seems as though Lovelock and his systems colleagues would like to erect systems theory as the all pervasive imperial ruler over the kingdom of science (and over the whole cascading empire of human knowledge).

To see the how the New Sciences are involved in such an intellectual colonising endeavour, witness this statement by one of its agents of colonisation in the area of organisational management:

I share the sentiments of physicist Frank Oppenheimer who says 'if one has a new way of thinking why not apply it wherever one's thought leads to?'[119]

By intellectual totalitarianism I am referring to the desire to have a theory explain the totality of existence (or a large chunk of it) without the need for alternative theories. This sort of totalitarianism is most manifest in what we might call the colonising of other disciplines. Systems theory, for instance, seeks to make itself the central idea in the study of biology, ecology, economics, society, physiology, etc. This type of colonising endeavour sometimes goes by the name of 'multi-disciplinarianism'. Systems theory is put forward as a theoretical nexus of various disciplines but if the traditions of those disciplines are inimical to the systems approach they are ignored; cast off as ephemeral practices, and consequently marginalised. This is the way that systems theory has characteristically rode into the various areas of its concern. In sociology it accepts the unity of culture of Spencer and Parsons without looking at the dissension of Marx. In ecology it looks at the integration of ecosystems and communities by Odum and Clements without looking at the disintegration of ecological communities by Gleason.

This intellectual totalitarianism started early in the history of what Postmodern Scientists would probably have us believe is the forerunner to their science; i.e. systems theory. For instance, when talking about the systems theorist Norbert Wiener, Fritjof Capra says:

> Wiener was not only a brilliant mathematician but also an articulate philosopher (in fact, his degree from Harvard was in philosophy). He was keenly interested in biology and appreciated the richness of natural, living systems. He looked beyond the mechanisms of communication and control to larger patterns of organisation and tried to relate his ideas to a wide range of social and cultural issues.[120]

Capra then goes on to showcase the universalist tendencies of another early systems theorist (and proto-Postmodern Scientist); Ludwig von Bertalanffy:

> since living systems span such a wide range of phenomena, involving individual organisms and their parts, social systems and ecosystems, Bertalanffy believed that a general systems theory would offer an ideal conceptual framework for unifying various scientific disciplines that had become isolated and fragmented.[121]

Capra also approvingly quotes Bertalanffy when the latter says:

> General Systems Theory should be an important means of controlling and instigating the transfer of principles from one field to another, and it will no longer

be necessary to duplicate or triplicate the discovery of the same principle in different fields isolated from each other.[122]

As already noted in an earlier chapter Bertalanffy himself has explained the exact nature of systems theory's comparative studies like this: "The search for such isomorphies is a major pursuit of systems science"; these isomorphies being:

a consequence of the fact that in certain aspects, corresponding abstractions and conceptual models can be applied to different phenomena. It is in view of these aspects that system laws will apply.[123]

Carrying on this tradition (that systems theorists might call multi-disciplinarianism but many postmodernist scholars might feel to be intellectual imperialism) the systems theorists of today also state that systems theory is applicable to diverse and seemingly disparate fields. Capra, for instance, states that since:

living systems cover such a wide range of phenomena--individual organisms, social systems, and ecosystems--the theory provides a common framework and language for biology, psychology, medicine, economics, ecology, and many other sciences, a framework in which the so urgently needed ecological perspective is explicitly manifest.[124]

Agreeing with Capra, and problematising it in terms of Postmodern Science, Kuppers claims that the new paradigm of Postmodern Science might have its roots in physics but has "applications to the humanities."[125]

Similarly Merry likes to see the New Sciences as being directly applicable to social science:

The social and behavioural sciences, both theoretical and applied, are at the beginning of the road of applying the insights of the New Sciences to their field of study and practice.[126]

Margeret Wheatley, as we have already seen, sees the complex systems theory of the New Sciences as being applicable to organisational management. Others, however, do not stop at proclaiming the migratory powers of systems ideas to just one particular discipline; they believe Postmodern Science's applicability is quite universal. Merry, for instance, talks of the 'order from chaos' concept (a central tenet in the New Sciences and also--as Fabel tells us--in Postmodern Science) as being applicable to many other areas. The 'order from chaos' concept, Merry says:

could be applied to chemical solutions, to termites building nests, to traffic jams and to the growth of cities. It could describe and explain transformation in a person's belief system, in organisations, in cultures, in political systems, and in all historical eras. In all of these systems new patterns of order emerge out of chaos. In all of these, the same phenomenon takes place.[127]

Mingers says much the same thing about another pet theory of Postmodern Science, self-organisation:

> Originally a biological concept [it] has made a remarkable impact not just on a single area, but across widely differing disciplines such as sociology, policy science, psychotherapy, cognitive science and law.[128]

Capra says much the same of self-organisation:

> Self-organisation...makes it possible to begin to understand biological, social, cultural and cosmic evolution in terms of the same pattern of systems dynamics.[129]

Davies also finds the universality of self-organisation appealing:

> Again and again we have seen examples of how organised behaviour has emerged unexpectedly and spontaneously from unpromising beginnings. In physics, chemistry, astronomy, geology, biology, computing--indeed, in every branch of science--the same propensity for self-organisation is apparent. [130]

Not to be outdone, Capra also reaffirms the total universality of self-organisation:

> As I mentioned before, living systems include individual organisms, social systems and ecosystems, and thus the new theory can provide a common framework and language for a wide range of disciplines--biology, psychology, medicine, economics, ecology and many others.[131]

Remembering the fondness with which Postmodern Science fans speak of Chaos theory (and also the intellectual closeness between Postmodern Science and New Science) we might note the glee with which New Scientist Alan Moore observes the ability of Chaos theory to explain everything:

> In a way Chaos and fractal maths gets rid of the need for a God. Previously people said that the unfathomable complexity of existence was the best argument for the existence of a creator. But Chaos and the Mandelbrot Set say that's not the case, that with one simple rule fed into a primordial mess, you can have infinitely complex, perfect order emerging.[132]

This universalism is such a strong current in complex systems theory that it has been expressed as being a core characteristic of the discipline. For instance witness Krohn *et al* who proclaim that we are:

> observing a present day scientific revolution encompassing many fields of the natural and social sciences as well as of the humanities.[133]

An explanation for the universal appeal of the systems theory within the New Sciences and Postmodern Science is attempted by Capra:

> it is useful to keep in mind that dynamical systems theory is not a theory of physical phenomena but a mathematical theory whose concepts and techniques apply to a broad range of phenomena. The same is true of Chaos theory and the theory of fractals, which are important branches of dynamical systems theory.[134]

Lilienfeld also talks about the mathematical abstraction of systems theory but he is not so positive. To Lilienfeld this mathematical abstractness is the prime heuristic and practical weakness of systems theory since:

> systems theory achieves its all encompassing 'universality' only by its very abstractness. All things are systems by virtue of ignoring the specific, the concrete, the substantive.[135]

By making this statement, Lilienfeld echoes the postmodern concern about abstract theory which throughout Modernity was allowed precedence over concrete stories of experience. In postmodernity, however, Capra's devotion to the abstractions of systems theory seem out of place.

Although fans of systems theory, the New Sciences and Postmodern Science constantly voice themselves as practitioners of holistic thought, their ongoing appeal to founding principles (like 'feedback', 'self-organisation', 'self-similarity' etc-- which all systems phenomena may be reduced to) exposes an entrenced reductionism within their work. If we define reductionism as does philosopher John Dupre: "Reductionism, in its broadest sense, is the commitment to any unifactorial explanation of a range of phenomena"[136], then we see that the founding principles within Postmodern Science are such examples of unifactorial explanation. All things; mountains, galaxies; ecosystems, traffic jams, eating habits etc., can be reduced to the mathematics and philosophy of negative feedback, fractal patterns and the 'order from chaos' idea.

So how does all this relate to postmodernism? The answer that academic scholars of postmodernism would give would surely be 'only negatively'. This sort of universalism is exactly what postmodernists object to. The drawing up of founding principles that can be applied to all phenomena is one of the most striking critiques of postmodernism:

> The postmodern mind does not expect any more to find the all-embracing, total and ultimate formula of life without ambiguity, risk, danger and error and is deeply suspicious of any voice that promises otherwise.[137]

Postmodernism talks about the pluralising and localising of knowledge and beliefs, not of their integration into one all encompassing, universalising and totalising set of ideas. If this is so then there is surely little justification for Postmodern Science to continue to call itself as such; since it is a name infected with self-contradiction.

We might like to ask, then, how Postmodern Scientists can possibly see themselves as postmodern? Having enunciated the varied and sometimes unconnected elements of postmodernism above, it is hard to see how comments like the following one from Birch can possibly be referring to postmodernism:

> Knowledge is lost in a sea of beliefs from a multitude of disciplines. The general purpose of the modern university is lost amid the incoherent variety of special purposes that have accumulated within it. The call of the postmodern worldview is for fewer beliefs and more belief.[138]

One may suspect that Birch and the other Postmodern Scientists are devoid of knowledge of postmodern theory but this is not the whole story. Birch, like Griffin, labels the postmodern negativity toward metanarratives 'deconstructive', whereas they themselves wish to pursue a 'constructive' brand of postmodernism where metanarratives are reassembled. Postmodern Science is thus, according to Gare[139] looking for a new 'mediating master discourse' produced by reflection and mediation between all other discourses.

This word, 'mediating', conjures up all sorts of connotations. It could refer to a mutually selected governor that brings to the fore a clarity of focus in assembling the concerns of differing players or it could mean the authoritative handing down of law that must be obeyed by both parties whether they like it or not. Most worryingly it conjures up the image of the latter proceeding under the guise of the former and this is the concern of deconstructive postmodernists. What might seem like an innocent, sharing, participatory, universally-pleasing, considerate and holistic framework for deciding the nature of the universe must always cavort dangerously with intellectual totalitarianism (where micro-narratives are marginalised if they conflict with larger metanarratives). No worldview can exist as an all-pervasive idea without the possibility of intellectual totalitarianism emerging from it.

Having said all of this, there are a few people that nevertheless put forward Postmodern Science as being capable of recognising the necessity of localised and pluralised metaphysical viewpoints. Wiessert, for example, when writing about Chaos theory, says:

> In non-linear dynamical systems, islands of order arise from the sea of chaos. The interspersed order comprehensible where chaos is not, implies abandoning the Modernist project of a global theory and attempting to compile a postmodern catalogue of local theories.[140]

While complex systems theory might be capable of recognising the localism of particular islands of order, all the islands of order arise according to the same laws as each other. They are separated by space but are nevertheless basically the same since they are thought by Postmodern Scientists and New Scientists to be governed by the same laws of formation. Thus, Wiessert's effort is but an uncommitted nod toward local theory and difference.

Such superficial recognition of localism and difference is noticeable throughout much of the writings on Postmodern Scienceand New Science. We can see it, for example, when we look at the 'levels of organisation' idea that New Scientists are so fond of explicating. The levels within such 'levels of organisation' are physically different from each other, hierarchically separated in space and time, but their underlying laws of formation and organisation remain constant. Thus any special local uniqueness of a level is outshone by the an overall metatheoretical commonality which applies to all levels.

Conclusion.

According to postmodern ideals the whole can only be dreamt of from within its parts; a broken assemblage of disintegrated complicated fragments that themselves often far exceed our ability to comprehend and imagine them. Postmodernists, as Bauman has indicated, have learnt to deal with this. Postmodern Science (which we might now see as indistinguishable from Modern science) is grossly uncomfortable with this inability to conjure up the wholes. Gaia is the result, a fantastic image of the whole Earth modeled in its entirety, complete with an ethics, a philosophy and a set of policy prescriptions. Chaos theory is also the result as it declares that those things once thought incomprehensibly disordered are now comprehensible and ordered.

Postmodern Science declares that the extirpation of Modernist principles from science must be undertaken in order to garner an environmentally friendly worldview. However Postmodern Science is itself unable to escape Modernism. This means that within its own intellectual framework, which states that an environmental worldview will only come about after Modernism has been ejected from science and natural philosophy, Postmodern Science might not be able to succeed in doing what it was designed to do.

If Postmodern Science does fail to eject Modernism from science, and if such an ejection is needed within science in order to let environmental values flourish, then what are we to do? Is there another way that science, or at least ecological science, might be postmodernised in a more thorough way so that it can properly call itself 'postmodern'. The answer to this question is revealed in Chapter Eight, where an alternative postmodern ecology is suggested. Before then, another problem that relates to the postmodernisation of the 'unity of nature' idea is explored.

Notes for Chapter 6 (What Is This Thing Called Postmodern Science?)

1. See, for example, T. Adorno and M. Horkheimer, *The Dialectic of Enlightenment* (NY: Cumming, 1947); J-F. Lyotard, *The Postmodern Condition* (Manchester: Manchester University Press); J. Habermas, *The Philosophical Discourse of Modernity* (Boston: Beacon Press, 1985); M. Foucault, "What is Enlightenment?", in P. Rabinow, ed, *The Foucault Reader* (Harmondsworth Penguin, 1986); S. Hall, S., ed, *Understanding Modern Societies* (Cambridge, Polity Press, 1992), R. Tester, *The Life and Times of Postmodernity* (London: Routledge, 1992); T. Docherty, ed, *Postmodernism: A Reader* (Hemel Hempstead: Harvester-Wheatsheaf, 1993).
2. See: D.R. Griffin, ed, *The Reenchantment of Science* (Albany: SUNY Press, 1988); C. Birch, *On Purpose* (Sydney: NSW University Press, 1990); C. Merchant, ed, *Key Concepts in Critical Theory: Ecology* (NJ: Humanities Press, 1994); D. Bohm, "Postmodern Science and a Postmodern World", in C. Merchant, ed, *Key Concepts in Critical Theory: Ecology* (NJ: Humanities Press, 1994) pp.342-350; A. Gare, *Postmodernism and the Environmental Crisis* (London: Routledge, 1995); F. Ferre, *Being and Value: Toward a Constructive Postmodern Metaphysical* (Albany: SUNY Press, 1996).
3. The postmodern movement when recounted by academics is usually found to make a critique over a whole spectrum of Modernist currents in history, society, art and culture. However, students of Postmodern Science tend to reduce Modernism to Modernist

Science, whose major drawbacks are held to be the scientific principles of reductionism, mechanicism, atomism and dualism. These are, of course, the same things that Birch, Ferre, Gare and others were critiquing before they latched onto the 'Postmodern Science' label (see Chapter One).

4. See R. Sassower, *Cultural Collisions: Postmodern Technoscience* (London: Routledge 1995), p.18. The use of definitions to elucidate upon an object, subject, act, term etc. would strike many postmodernists as a way of not only describing what that object, subject, act or term *is* but also what it *isn't*. Definitions they would say, categorically set up and describe what something is capable of being and what something is incapable of being. Definitions also enforce one dimensional elaboration of things that might potentially be explained in a myriad of ways if these ways were not categorically excluded by the definition.
5. I. Hassan, *The Postmodern Turn: Essays in Postmodern Theory and Culture* (Ohio: Ohio State University Press, 1987) p.168.
6. The quotes are taken from, respectively: Hassan, *The Postmodern Turn*, p.167; A. Heller, and F. Feher, *The Postmodern Political Condition* (Cambridge: Polity Press, 1988) p.1; P. Zagorin, "Historiography and Postmodernism: Reconsideration", *History and Theory*, 29, (1990) p.265.
7. C. West, "The New Cultural Politics of Difference", in S. Seidman, ed, *The Postmodern Turn* (London: Routledge, 1994), 65-81.
8. See, for instance: H.F. Haber, *Beyond Postmodern Politics* (NY: Routledge, 1994).
9. P. Rosenau, *Postmodernism and the Social Sciences* (Princeton: Princeton University Press, 1992) p.3.
10: Tester, *The Life and Times of Postmodernity*, p.28.
11. *ibid.*
12. N. Fraser and L. Nicholson, "Social Criticism Without Philosophy: an Encounter between Feminism and Postmodernism", in S. Seidman, ed, *The Postmodern Turn: New Perspectives on Social Theory* (Cambridge University Press, Cambridge, 1994) p.139.
13. The book by Donald Griffin is D.R. Griffin, D.R., ed, *The Reenchantment of Science* (Albany: SUNY Press, 1988). This forms, according to Griffin's introduction, the first installment on a series on 'Constructive Postmodern Science'.
14. See Griffin's introduction to: Griffin, ed, *The Reenchantment of Science*, p.x.
15. A.J. Fabel, "Environmental Ethics and the Question of Cosmic Purpose", *Environmental Ethics*, 16, (1994) p.303.
16. A. Gare, *Postmodernism and the Environmental Crisis* (London: Routledge, 1995), p.110.
17. C. Birch, *On Purpose* (Sydney: UNSW Press, 1990) p.75.
18. *ibid*, p.128.
19. *ibid*, p.140.
20. See N.K. Hayles, *Chaos Bound: Orderly Disorder in Contemporary Literature and Science* (Ithaca: Cornell University Press, 1990); T.P. Wiessert, "Representation and Bifurcation: Borge's Garden of Chaos Dynamics", in N.K. Hayles, ed, *Chaos and Order: Complex Dynamics in Literature and Order*, (Chicago: Chicago University Press, 1991) pp.234-250.

21. S. Best, "Chaos and Entropy: Metaphors in Postmodern Science and Social Theory", *Science as Culture*, 2, 11 (1991) p189. The theme of this paper is reiterated in: S. Best and D. Kellner, *The Postmodern Turn* (New York: Guilford Press, 1997)
22. *ibid.*
23. D. Bohm, "Postmodern Science and a Postmodern World", in D.R. Griffin, ed, *The Reenchantment of Science* (Albany: SUNY Press), pp.57-68.
24. See, for example: F. Ferre, "Religious World Modeling and Postmodern", in D.R. Griffin, ed, *The Reenchantment of Science* (Albany: SUNY Press, 1988) pp.87-98; F. Ferre, "Obstacles on the Path to Organismic Ethics: Some Second Thoughts", *Environmental Ethics*, 11, (1989): 231-241; F. Ferre, *Hellfire and Lightning Rods: Liberating Science, Technology and Religion* (NY: Orbis Books, 1993); F. Ferre, *Being and Value: Toward a Constructive Postmodern Metaphysical* (Albany: SUNY Press, 1996).
25. S. Krippner, "The Holistic Paradigm", *World Futures*, 30, (1991) :133.
26. Gare, *Postmodernism and the Environmental Crisis*, p.3.
27. *ibid*, p.87.
28. Best, "Chaos and Entropy: Metaphors in Postmodern and Social Theory" p.189.
29. There is also another important purpose in the laying out of this section. Later in this book (i.e. in Chapter Eight: An (Other) Postmodern Ecology) I enter into an altogether different attempt to postmodernise ecological science, natural philosophy and environmentalism. This alternative attempt at a postmodernising effort will also draw on some of the ideas outlined in this section. This dual purpose must be kept in mind throughout the reading of this section. What must also be noted when viewing this section is the fact that it is not me characterizing the following listed attributes as postmodern but the many people cited and named with the various attributes.
30. For instance, as expressed in: J-F. Lyotard, *The Postmodern Condition* (Manchester: Manchester University Press, 1984); T. Eagleton, *Against the Grain*, (London: Verso, 1986); D. Harvey, *The Condition of Postmodernity* (Oxford: Blackwell, 1990); Best, "Chaos and Entropy: Metaphors in Postmodern Science and Social Theory"; S. Grenz, *A Primer on Postmodernism* (Grand Rapids: Eerdmans, 1996); I.H. Grant, "Postmodernism and Science and Technology", in S. Sim, ed, *Postmodern Thought* (Oxford: Icon, 1998) pp.65-77.
31. Grenz, *A Primer on Postmodernism*, p.46.
32. Eagleton quoted in Harvey, *The Condition of Postmodernity*, p.9.
33. Best, "Chaos and Entropy: Metaphors in Postmodern Science and Social Theory", A. Kroker and D. Cook, *The Postmodern Scene: Excremental Culture and Hyper-Aesthetics* (London: MacMillan, 1991), Z. Bauman, *Postmodern Ethics* (MA: Blackwell, 1993), Grenz, *A Primer on Postmodernism*.
34. Z. Bauman, *Postmodern Ethics*, p.32.
35. Best, "Chaos and Entropy: Metaphors in Postmodern and Social Theory", p.213.
36. P. Davies, *The Cosmic Blueprint* (Harmondsworth: Penguin: 1987) p.197.
37. For instance, as expressed in: J. Baudrillard, *Simulacra and Simulations* (NY: Semiotext(e), 1983); L. Appignanesi and H. Lawson, *Dismantling the Truth: Reality in a Postmodern World* (NY: St. Martins Press, 1989); Z. Bauman, *Intimations of Postmodernity* (London: Routledge, 1992); Rosenau, *Postmodernism and the Social*

Sciences; M.E. Zimmerman, *Contesting Earth's Future: Radical Ecology and PostModernity* (Berkeley: University of California Press, 1994); A. Easthope, "Postmodernism and Critical and Cultural Theory", in S. Sim, ed, *Postmodern Thought* (Cambridge: Icon Books, 1998) pp.15-27; D. Scott, "Postmodernism and Music", in S. Sims, ed, *Postmodern Thought* (Cambridge: Icon Books, 1998) pp.134-146.

38. For instance, as expressed in: G. Vattimo, *The End of Modernity* (Cambridge: Polity Press, 1988); Zimmerman, *Contesting Earth's Future*; C. Keolb, ed, *Nietzsche as Postmodernist: Essays Pro and Contra* (Albany: SUNY Press, 1996).
39. Nietzsche's view on truth may be summarised by quoting the following well-known passage:

> "What, then, is truth? A mobile army of metaphors, metonymies, anthropomorphisms, a sum, in short, of human relationships which, rhetorically and poetically intensified, ornamented, and transformed come to be thought of , after long usage by a people, as fixed, binding and canonical. Truths are illusions which we have forgotten are illusions, worn out metaphors, now impotent to stir the senses, coins which have lost their faces and are now considered as metal rather than currency." (taken from F. Nietzsche, "Philosophy and Truth", in D. Breazeal, ed, *Selections from Nietzsche's Notebooks of the Early 1870s* (NJ: Humanities Press, 1979) p.84.

40. Rosenau, *Postmodernism and the Social Sciences*, p.31.
41. See for example B. Latour and S. Woolgar, *Laboratory Life: The Construction of Scientific Facts* (Princeton: Princeton University Press, 1979); B. Latour, *Science in Action: How to Follow Scientists and Engineers Through Society* (Cambridge: Harvard University Press; 1988).
42. Rosenau, *Postmodernism and the Social Sciences*, p.110.
43. Zimmerman, *Contesting Earth's Future*, p.94.
44. Rosenau, *Postmodernism and the Social Sciences*, p.78.
45. Zimmerman, *Contesting Earth's Future*, p.135.
46. As stated in Rosenau, *Postmodernism and the Social Sciences*, p.111.
47. Zimmerman, *Contesting Earth's Future*, p.100.
48. For instance, although there are hundreds of disciplines and sub-disciplines in biology all are believed to have come to terms with the founding idea of 'evolution via natural selection'. Indeed, they must do this in order to sustain their scientificity. In this way the founding idea of evolution is not merely a temporary truth which might be abandoned if a better founding principle comes along but is actually an unrelinquishable premise which defines scientificity and truthhood. There may be little things wrong with the theory of evolution by natural selection, the argument of Modern scientists would go, but these things will be 'sorted out through the course of good science'. This means that the 'evolution via natural selection' theory as a whole is thereby ossified into biology as nothing other than an obvious truth
49. A conclusion that Don Miller is also forced to arrive at: D. Miller, "Metaphor, Thinking and Thought", *ETC.: A Review of General Semantics*, 39, (1982) :134-50; D. Miller, "Metaphor, Thinking and Thought: Part Two", *ETC.: A Review of General Semantics*,

39, (1982) :242-256; D. Miller, "Metaphor and Culture", *Paper Presented to the 1983 Culture Seminar*, University of Melbourne, Australia. (1983).

50. G. Lakoff, and M. Johnson, *Metaphors We Live By* (Chicago: University of Chicago Press, 1980) p.3.
51. *ibid*, p.160
52. Miller, Metaphor and Culture, 6.
53. Davies, *The Cosmic Blueprint*, p.203.
54. For instance, as expressed in: Hassan, *The Postmodern Turn*; Bauman, *Intimations of Postmodernity*; Rosenau, *Postmodernism and the Social Sciences*.
55. Bauman, *Intimations of Postmodernity*, p.29.
56. Rosenau, *Postmodernism and the Social Sciences*, p.14.
57. For instance, as expressed in: J. Derrida, *Of Grammatology* (Baltimore: Johns Hopkins University Press, 1976); Bauman, *Intimations of Postmodernity*.
58. See Birch, *On Purpose*, pp.81-85.
59. R. Barthes, *S/Z* (Oxford: Blackwell, 1975); Derrida, *Of Grammatology*, J-F. Lyotard, *The Differend: Phrases in Dispute* (Manchester: Manchester University Press, 1988); Kroker and Cook, *The Postmodern Scene*.
60. For instance, as expressed in: Lyotard, *The Postmodern Condition*; M. Poster, *The Mode of Information: Poststructuralism and Social Context* (Chicago: Chicago University Press, 1990); D.R. White, *Postmodern Ecology: Communication, Evolution and Play* (Albany: SUNY Press, 1997).
61. See Miller, "Metaphor, Thinking and Thought"
62. S. Sim, Postmodernism and Philosophy, in S. Sim, ed, *Postmodern Thought* (Cambridge: Icon, 1998) :316.
63. W. Steiner, "Language as Process: Serge Karchevskij's Semiotics of Language", in L. Matejka Sound, ed, *Sign and Meaning* (Michigan: University of Michigan Press, 1976).
64. Or so say Kroker and Cook, *The Postmodern Scene*, p.vii.
65. R. Lilienfeld, *The Rise of Systems Theory: An Ideological Analysis* (Krieger, 1988).
66. For instance, as expressed in: C. Jencks, *What is Postmodernism?* (London: Academy Edition, 1986); Harvey, *The Condition of Postmodernity*; Rosenau, *Postmodernism and the Social Sciences*; R. Appignanesi *et al*, *Postmodernism for Beginners* (Cambridge: Icon, 1995); B. Lewis, "Postmodernism and Literature", in S. Sims, ed, *Postmodern Thought*, (Cambridge: Icon Books, 1998) pp.121-132.
67. For instance, as expressed in: Heller and Feher, *The Postmodern Political Condition*; Harvey, *The Condition of Postmodernity*; D. Ashley, "Playing with the Pieces: The Fragmentation of Social Theory", in P. Wexler, ed, *Critical Theory Now* (London: Falmer Press, 1991) pp.70-97; Lewis, "Postmodernism and Literature".
68. Bohm, "Postmodern Science and a Postmodern World, p.86.
69. Harvey, *The Condition of Postmodernity*, p.44.
70. For instance, as expressed in: Lyotard, *The Differend*; S.K.White, *Political Theory and Postmodernism* (Cambridge: Cambridge University Press, 1991).
71. J. Lovelock, *The Ages of Gaia*, (NY: Norton & Co, 1988) p.xvii.
72. Rosenau, *Postmodernism and the Social Sciences*, p.97.

73. Harvey, *The Condition of Postmodernity*; White, *Political Theory and Postmodernism*; Tester, *The Life and Times of Postmodernity*.
74. Harvey, *The Condition of Postmodernity*, p.47.
75. Barthes, *S/Z*; M. Riffaterre, *Text Production* (NY: Columbia University Press, 1983); Rosenau, *Postmodernism and the Social Sciences*.
76. Harvey, *The Condition of Postmodernity*, p.49.
77. S. Heath, "Intertextuality", in M. Payne, ed, *A Dictionary of Cultural and Critical Theory* (Oxford: Blackwell, 1998) p.258.
78. Rosenau, *Postmodernism and the Social Sciences*, p.xii.
79. *ibid*, p.xiv.
80. Birch, *On Purpose*.
81. Bohm, "Postmodern Science and a Postmodern World", p.347.
82. Wiessert, "Representation and Bifurcation".
83. Harvey, *The Condition of Postmodernity*; Hassan, *The Postmodern Turn*; Grant, "Postmodernism and Science and Technology"; Lewis, "Postmodernism and Literature".
84. Best, "Chaos and Entropy".
85. Wiessert, "Representation and Bifurcation". Likewise Chaos theory is associated with postmodernism in: T.R. Young, "Chaos Theory and Symbolic Interaction Theory: Poetics for the Post Modern Sociologist", *Symbolic Interaction*, 14, (1991) :321-334; Hayles, *Chaos Bound*; D. Worster, "The Ecology of Order and Chaos" in S.J. Armstrong and R.C. Botzler, eds, *Environmental Ethics: Divergence and Convergence* (NY: McGraw-Hill, 1993) pp.39-48; Zimmerman, *Contesting Earth's Future*; Gare, *Postmodernism and the Environmental Crisis*; D. Wilcox, "What does Chaos theory have to do with art?", *Modern Drama*, 39, 4, (1996) :698-711; S. Kember, "Feminist Figuration and the Question of Origin", in G. Robertson *et al*, eds, *FutureNatural* (London: Routledge, 1996) p.256-269, T. Jagtenberg, and S. McKie, *Eco-Impacts and the Greening of Postmodernity* (CA: Sage, 1997); D. Milovanovic, *Postmodern Criminology* (New York: Garland, 1997); P. Cilliers, *Complexity and Postmodernism: Understanding Complex Systems*, London: Routledge, 1998); Grant, "Postmodernism and Science and Technology". W. Rasch *et al*, eds, *Observing Complexity: Systems Theory and Postmodernity* (Minneapolis: University of Minnesota, Press, 2000); G. Slethaug, *Beautiful Chaos: Chaos Theory and Metachaotics in Recent American Fiction* (Albany: SUNY Press, 2000).
86. Best, "Chaos and Entropy" p.202.
87. Best, "Chaos and Entropy".
88. U. Merry, *Coping With Uncertainty: Insights from the New Sciences of Chaos, Self-Organisation and Complexity* (Westport: Praeger, 1995) p.13.
89. Rosenau, *Postmodernism and the Social Sciences*, p.170.
90. See the introduction in N.K. Hayles, ed, *Chaos and Order: Complex Dynamics in Literature and Science* (Chicago: Chicago University Press, 1991), p.4.
91. Merry, *Coping With Uncertainty*, p. 53.
92. Bauman, *Intimations of Postmodernity*; B.E. Babich *et al*, "On the Idea of Continental and Postmodern Perspectives in the Philosophy of Science", in B.E. Babich *et al*, eds,

Continental and Postmodern Perspectives in the Philosophy of Science (Aldershot: Avebury, 1995) pp.1-10.

93. Babich *et al*, "On the Idea of Continental and Postmodern Perspectives in the Philosophy of Science" p.1.
94. E. Goldsmith, *The Way: An Ecological Worldview* (NY: Shambhala, 1993), p.xvii.
95. Davies, *The Cosmic Blueprint*.
96. J.O. Slack and L.A. Whitt "Ethics and Cultural Studies", in L. Grossberg *et al*, eds, *Cultural Studies* (NY: Routledge, 1992), pp.571-592.
97. Davies, *The Cosmic Blueprint*, p.132.
98. Birch, *On Purpose*.
99. Grenz, *A Primer on Postmodernism*, p.7.
100. J. Farganis, *Readings in Social Theory: From the Classic Tradition to Postmodernism* (NY: McGraw-Hill, 1993), p.364.
101. Hassan, *The Postmodern Turn*; Heller and Feher, *The Postmodern Political Condition*; Zagorin, "Historiography and Postmodernism: Reconsideration", Zagorin, 1990; Sim, ed, *Postmodern Thought*.
102. M. Foucault, 'Preface' to G. Deleuze & F. Guattari, *Anti-Oedipus*, (Minneapolis: University of Minnesota Press, 1983), p.xiii.
103. Hassan, *The Postmodern Turn*; J. Deleuze and F. Guattari, *A Thousand Plateaus* (Minneapolis: Minnesota University Press, 1987); Bauman, *Intimations of Postmodernity*.
104. Bauman, *Intimations of Postmodernity*, p.34.
105. Deleuze and Guattari, *A Thousand Plateaus*.
106. Hassan, *The Postmodern Turn*, Harvey, *The Condition of Postmodernity*; Bauman, *Intimations of Postmodernity*.
107. H. Skolomowski, *Eco-philosophy: Designing New Tactics for Living* (Boston: Boyars, 1991), p.100.
108. Birch, *On Purpose*, p.43.
109. See, as examples from the 1990s, M. Bookchin, *The Philosophy of Social Ecology: Essays in Dialectical Naturalism* (Montreal: Black Rose, 1990).; Goldsmith, *The Way*, and R. Sylvan, "Illusion and Illogic in Evolution" *Revista Di Biologica*, 87, (1994) 191-221.
110. Bauman, *Intimations of Postmodernity*, p. 35.
111. M. Wheatley, *Leadership and the New Science: Learning About Organisation from an Orderly Universe* (San Francisco: Berret-Koehler Publishers, 1992), p.187.
112. Hassan, *The Postmodern Turn*; Harvey, *The Condition of Postmodernity*; S. Best and D. Kellner, *Postmodern Theory: Critical Investigations* (London: MacMillan, 1991); Lyotard, *The Differend*.
113. Best and Kellner, *Postmodern Theory*, p.4.
114. K.M. Haney, *Intersubjectivity Revisited: Phenomenology and the Other*, (Athens: Ohio University Press, 1994).
115. Bauman, *Intimations of Postmodernity*; Lyotard, *The Postmodern Condition*; M. Foucault, "Geneology and Social Criticism", in S. Seidman, ed, *The Postmodern Turn:*

New Perspectives on Social Theory, (Cambridge: Cambridge University Press, 1994), pp.39-45; Sim, ed, Postmodern Thought.
116. Bauman, *Intimations of Postmodernity*, p.39.
117. *ibid*, p.190.
118. J. Lovelock, *Healing Gaia: Practical Medicine for the Planet* (NY: Harmony, 1991) p.90.
119. Wheatley, *Leadership and the New Science*, p.13.
120. Capra, *The Web of Life*, p.54.
121. *ibid*, p.116.
122. *ibid*, p.49.
123. L.v. Bertalanffy, General System Theory, in N.J. Demerath & R.A. Peterson, eds, *System, Change and Conflict* (NY: Free Press, 1967), p.118.
124. F. Capra, Systems Theory and the New Paradigm, in C. Merchant, ed, *Key Concepts Critical Theory: Ecology* (NJ: Humanities Press, 1994), p.335.
125. B-O. Kuppers, "On a Fundamental Shift in the Natural Sciences", in W. Krohn et al, eds, *Self-Organisation: Portrait of a Scientific Revolution*, Dordrecht: Kluwer Academic, 1990), p.52.
126. Merry, *Coping With Uncertainty*, p.24.
127. *ibid*, p.44. See also: I. Prigogine and I. Stengers, *Order Out of Chaos* (NY: Bantom Books, 1984); A.J. Fabel "Environmental Ethics and the Question of Cosmic Purpose", *Environmental Ethics*, 16, (1994) pp.303-314.
128. J. Mingers, *Self-Producing Systems* (NY: Plenum, 1995), p.*ix*. See also H. Maturana and F. Varela, *Autopoiesis and Cognition* (D. Reidel, 1980)
129. F. Capra, *The Turning Point* (NY: Simon & Schuster, 1982), p.310. See also E. Jantsch, *The Self-Organising Universe* (Oxford: Pergamon Press, 1980)
130. Davies, *The Cosmic Blueprint*, p.200.
131. F. Capra, Systems Theory and the New Paradigm, in C. Merchant, ed, *Key Concepts Critical Theory: Ecology* (NJ: Humanities Press, 1994), p.338.
132. Moore quoted in R. Wright, "Art and Science in Chaos: Contested Readings of Scientific Visualization", in R. Robertson et al, eds, *FutureNatural* (London: Routledge, 1996), p.228.
133. Some notable examples of such migration of the New Sciences into various disciplines already occurring include: literary theory (Hayles, *Chaos Bound*), psychology (W. Sulis and A. Combs, eds, *Nonlinear Dynamics in Human Behavior*, NJ: World Scientific, 1996), planning (T.J. Cartwright, "Planning and Chaos Theory", *APA Journal*, Winter 1991, pp.44-56), art theory (V. Sobchack, "A Theory of Everything: Meditations on Total Chaos", *Artforum*, November 1990, pp.148-55; L. Schlain, *Art and Physics: Parallel Visions in Space, Time and Light*, NY: Morrow, 1991), economics and business (D. Parker and R. Stacey, *Chaos, Management and Economics*, Sydney: Centre for Independent Studies, 1995; M. Anderla *et al*, *Chaotics: An Agenda for Business and Society in the 21st Century*, Twickenham: Adamantine Press, 1997), geology (D. Turcotte, *Fractals and Chaos in Geology and Geophysics*, NY: Cambridge University Press, 1992), population geography (D.S. Dendrinos, *Chaos and Socio-spatial Dynamics*, NY: Springer-Verlag, 1990), neurobiology (E. Basar, ed, *Chaos in Brain Function*, Berlin, Springer-Verlag, 1990), history (R. Abraham, *Chaos, Gaia:*

Eros: A Chaos Pioneer Uncovers the Three Great Streams of History, San Francisco: Harper, 1994), electronics (M. Van Wyk and W. Steeb, W. *Chaos in Electronics*, Boston: Kluwer, 1997); organisational studies (Wheatley, *Leadership and the New Science; Parker and Stacey, Chaos, Management and Economics*); physiology, (J.B. Bassingthwaighte *et al*, *Fractal Physiology*, Oxford, Oxford University Press, 1994). It might be noted here that given its tendency to cross-disciplines, postmodernism might itself be accused of being a colonising discourse. However, most postmodernists would hold that postmodernism hardly issues programmatic statements aimed to inspire trans-disciplinary colonisation. Instead postmodernists would point out that when postmodernism does cross disciplines it does so in an attempt to re-story the human sciences so that new stories are opened up or made possible. Systems theory, on the other hand, closes down story-telling by issuing a monolithic, universal story that acts to position all other stories within itself. Where postmodernism attempts to pluralise stories wherever it goes, systems theory--and its relations--act to unify various stories. This difference can be demonstrated by acknowledging that the stories of systems theory can all be related to the key concepts of self-organisation and feedback. When systems theory (and New Science and Postmodern Science) advance narratives about anything in nature they always include the essential processes of self-organisation or feedback. Without them, systems theory would have no story-telling potential; it would not exist. On the other hand, there is no essence to postmodernism; no essential focal narrative which must be retold in order to characterise a postmodern story as postmodern. Postmodernism is more a 'structure of feeling' (see F. Pfeil, "Postmodernism as a Structure of Feeling", in L. Grossberg *et al*, eds, *Cultural Studies*, NY: Routledge, 1988, pp.592-599) than a codified theoretical framework. While systems theory's identity is very strongly linked with its essence of self organisation and feedback, postmodernism has no identity; or if it does it is plural and heterogeneous. This lack of a central plank means that postmodernism, when it does enter into unfamiliar intellectual territory, can be moulded and shaped at will by those whose territory it is entering.

134. Capra, *The Web of Life*, p.112.
135. R. Lilienfeld, *The Rise of Systems Theory: An Ideological Analysis*, (Krieger, 1988), pp.192.
136. J. Dupre, *Disorder of Things: Metaphysical Foundations of the Disunity of Science* (Cambridge: Harvard University Press, 1993) p.87.
137. Bauman, *Postmodern Ethics*, p.245.
138. Birch, *On Purpose*, p.140.
139. Gare, *Postmodernism and the Environmental Crisis*, p.159.
140. Wiessert, "Representation and Bifurcation", p.225.

CHAPTER 7

Mechanicism Vs Organicism: A False Dichotomy?

Introduction

Modern science is mechanistic and the worldview that such science gives rise to is a mechanised worldview; or so say many of the environmental and scientific thinkers talked about in previous chapters.[1] According to Moore:

> this mechanisation of the world has led us to believe that we live in an essentially dead universe where life arises only by chance. The result has been a total separation of ourselves from nature. This separation finally reached its logical extreme in the modern myth...that we can use technology to dominate, manage and control nature for our own benefit.[2]

There is much claim by Postmodern Scientists that what we need in science is an initiation or resurrection of organic thinking in order to make a sustained attack on mechanistic thinking. This organicism is what Postmodern Scientists believe lies in Postmodern Science. It is also what New Scientists believe lies in the New Sciences, what holistic philosophers of nature believe lies in ecological holism, and what systems thinkers believe lies in systems science. All these 'organic' sciences, together, tend to believe that the Modernist evil of mechanicism is rejected and ejected from their new forms of thinking to reveal the world--and the things in it--as actively creative agents intimately connected to one another.

By encouraging the supersession of mechanicism by organicism, Postmodern Scientists, New Scientists, holist environmentalists etc, are making a declaration that Clements should replace Gleason, that Gaian systems theory should replace the fragmented sub-disciplines of biology and ecology, and that self-organisation should become a prime theory to help explain life, the universe and everything in it.[3]

There are problems with this angle of attack. Notably, it is not always clear, historically, where mechanicism and organicism arise from. Nor is it easy to distinguish between the two from a philosophical perspective.

The Metaphysics of Organicism and Mechanicism

So superorganismic ecology, the Gaia theory and complexity theory might be cited as organic. What does that mean? Apart from the obvious idea that "an organic worldview generalises from observed organisms"[4] instead of machines, the idea of organicism is often thought to be closely tied to the concept of unity. Witness this definition of what it is to be organic:

> There are two pervasive views of nature which were evident in traditional natural history and persist in current discussions of ecological theory:
> 1) Mechanical = actions of individual parts of a whole are explained by known laws, and the whole is the sum of the parts and their interactions.
> 2) Organic = The whole exists first and its design explains the actions of the parts.[5]

For those who are in the business of promoting organicism, as well as for many who are not, mechanicism is allied to materialism, reductionism, and atomism. For instance, Ted Benton states that whether you may call it mechanical reductionism, reductionist materialism or mechanical materialism, mechanicism "asserts that living organisms are subject to the same laws as those that govern inorganic nature."[6]

David Bohm, likewise, links mechanicism to reductionism by saying:

> The first point about mechanicism is that the world is reduced as far as possible to a set of basic elements...Second, these elements are basically external to each other...because the elements only interact mechanically by sort of pushing each other around, the forces of interaction do not affect their inner natures.[7]

Similarly Charles Birch contends:

> The mechanistic model is properly called atomistic...Its method consists of subdividing the world into its smallest parts, which at one time were thought to be atoms. The essence of atomism is that these parts remain unchanged no matter what particular whole they constitute, be it a stone or a brain. Having divided the universe into its smallest bits you then try to build it up again, and of course when you do you get a machine.[8]

Similarly, Cobb sees that mechanicism is equivalent to the materialism of what he calls 'substantialism': "a substance is that which depends on nothing else for its existence" and under mechanicism such "a thing remains fundamentally the same regardless of its relations."[9]

To counteract mechanicism (and in contrast to materialism, atomism and reductionism) both Birch and Cobb advocate what they regard as the postmodern alternatives: 'processes over substances', 'events over objects', 'wholes over parts'; in short 'organicism over mechanicism'. For both Birch and Cobb, as well as for Capra, Davies and Bohm, the shift from mechanical thought to holistic process thought is adequately encapsulated in systems theory and complex systems theory. The closeness of the organic approach to systems thinking, as interpreted by organicists, is reflected in Capra's assertion that:

> It is perhaps worthwhile to summarise the key characteristics of systems thinking at this point. The first, and most general, criterion is the shift from the parts to the whole...In the shift from mechanistic thinking to systems thinking, the relationship between the parts and the whole is reversed. Cartesian science believed that in any complex system, the behaviour of the whole could be analysed in terms of the properties of the parts. Systems science shows that living systems cannot be understood by analysis.[10]

So whereas "the mechanistic outlook reduces reality to a set of basic elements or elementary particles and forces"[11], the organic outlook sees every element as 'interconnected', 'interrelated', 'interdependent', and 'part of a greater whole'.

From such prescriptions, it is easy to observe here the division thought to lie between some of the ecological traditions and environmental concepts that have been talked about so far in this book. Clements, Lovelock and Odum are organic (and indeed they have all had the word attached to their work) while Gleason and the community ecologists are mechanical. Unfortunately, though, this dichotomy does not hold up in practice, such that (in ecology at least) mechanicism and organicism is probably a false dichotomy.

One way to show that organicism and mechanicism are hardly appropriate tools by which to dichotomise ecological science is to examine those instances in which traditional mechanistic ecology can be shown to possess what organicists themselves class as 'organic' principles.

In the substance thinking of mechanicism "the substance is independent of relations and then enters into relations which are always external ones."[12] Bohm expresses this idea thus:

> these elements are basically external to each other; not only are they separate in space, but even more important the fundamental nature of each is independent of

> that of the other...Third, because the elements only interact mechanically by sort of pushing each other around, the forces of interaction do not affect their inner natures.[13]

This is contrasted against the process thinking of systems theory whereby the primary entities of existence, i.e. events (instead of substances), are fully constituted by their interactions. In physics this means that a particle such as an electron turns from being a mechanical 'substance' (that collides into others while not being affected by the collision except so far as its position in space changes) into being an event (which is intimately affected by the other events that go on around it). We can see how this dichotomy might be applied in ecopolitical philosophies of nature since in an organic ecological worldview the various constituents of nature, both living and non-living, are turned from mechanical ecological atoms that collide with each other without really affecting one another, into organic events that intimately change the nature of each other when they meet. This is to say the major critique that organicist unitarians would level at atomistic science (including atomistic ecology) is that it treats the members of the living world as though they were mere mechanical substances.

This might be an adequate criticism of substantialism in ecology if it actually concurred with what non-holist ecologists (for example, the Gleasonians) say about what happens in the world but it does not. The substances that non-holist ecologists talk about are not conceived of as ecological atoms but are conceived of as enormously heterogeneous interacting beings capable of having their 'internal structure' (i.e. their actions and their characteristics) changed when they collide with other 'substances'.

In fact, with regards to ecology, it seems as though the criticism emerging from organicism might be more applicable to process thinking. This is to say we can easily imagine in process ecology (which emphasises 'events' over 'substances') events which move around and interact with each other yet remain unchanged by the interaction. This is, for example, the point that is made about ecosystems ecology (the preferred ecology of organicists) in Chapter Two, since the events in ecosystem ecology (organismal growth, cycling flows, material transfer, biomass production) might interact profusely but still give rise to unchanging essences (i.e. elemental matter and energy) that remain the same whatever interactions they undertake. Thus it seems, at least in ecology, you can have mechanical events just as you can have mechanical substances and you can have organic substances just as you can have organic events.

Another way to deconstruct the dichotomy of the organic versus the mechanical is to look at the intellectual heritage of organic thought. Here we find some telling dilemmas.

The Intellectual Heritage of Organic Thought

When taking a look at the relevance of organic versus mechanical debates as they relate to the ecological world, it is important to focus on that organic entity which organic ecologists are so proud of; the one which is so indebted to holism, the one which is so guided by the physical laws of the New Sciences, and the one which is the basis of environmental policy-making; the ecosystem.

As has already been explored[14], the usual story in the history of ecology is that the ecosystem was a concept dreamt up by Arthur Tansley in the 1930s yet did not really make any difference in ecological practice until the likes of Raymond Lindemanused it as a methodological approach in his studies of actual geographical sites in the 1940s. Furthermore, the ecosystem concept as an environmentally-relevant unit did not really take-off until Eugene Odum and his brother Howard popularised it, published a textbook about it, and gained grant money for studying it, in the 1950s and 60s.

For Eugene Odum and his followers the ecosystem was talked about as being the fundamental unit of ecology. This is a view taken up and supported in modern environmental practice and policy-making and also in much environmentally-focused philosophy. Although the tradition in this environmentally-focused philosophy is to believe that ecosystems operate holistically, i.e.: organically, such beliefs are somewhat ironic given the desire by the ecosystem concept's originator, Arthur Tansley, to have it used as a mechanising influence in ecology. It is also strange given that ecosystem studies mainly focus on physical and not biological aspects of natural areas. I might restate here Golley's comments about systems-analysed species being mere mechanical objects when he says:

> Although the advantages were many, the disadvantage was that most of biological reality encompassed in the species was lost. In the ecosystem model, species acted abstractly, like robots. The decision to cut ecosystems studies off from biology and natural history linked them more closely to engineering, physics and mathematics.[15]

Despite this, the writings of the Odums and other ecosystem ecologists are routinely utilised to promote organicism. Edward Goldsmith, for example, praises the organic viewpoint of Odum's ecology for its inherent organicism only to go on to unwittingly support mechanicism by favourably quoting Eugene Odum when Odum used intensely mechanical metaphors. For instance, in *The Way* Goldsmith firstly praises Odum:

> Eugene Odum, one of the few remaining academic ecologists whose work has not been perverted to fit the paradigm of mechanistic science.[16]

Then Goldsmith berates modern ecology for being too quantifiable, too obsessed with energy and for being:

> reconcilable with the paradigm of science which sees living things as no more than machines.[17]

And finally Goldsmith approvingly quotes Odum when Odum himself is working within the paradigm of science Goldsmith has just denounced:

> Eugene Odum notes how ecosystems are endowed with the necessary mechanisms for self-regulation and hence homeostasis. These information networks Odum thought as 'the invisible wires of nature' or alternatively as the 'hormones of ecosystems'.[18]

In a similar vein, Stan Rowe makes the same sort of self-contradictory case for organic ecology.[19] First, Rowe decries the reductionism of modern ecology for concentrating too much on the parts of nature rather than nature as a whole and he asks for us to reconsider Odumian systems ecology as the scientific manifestation of environmental thought. Eugene Odum's work in the 1950s and 60s, Rowe tries to make us believe, is a manifest example of the holist approach that environmentalists are desiring in science. The 1950s and 1960s work that Rowe identifies as being of an organic nature, however, is the same work in which Odum charts out the similarities of ecosystem flow to 'networks', 'circuits', 'organismal physiology', 'machines', 'water-mills' and 'automobiles', which are all very strong mechanical analogies.

Equally ambiguous to organic ecological sensibilities in this respect is the fact that the Odums, and other ecologists that followed them, used mixed organic and mechanical metaphors when describing their 'fundamental unit of ecology', the ecosystem. According to Hagen:

> the Odums' writings exemplify this Janus-like conception of ecological systems...For example when he needs an analogy for communication and control mechanisms in the ecosystem, Eugene turns to the physiology of the endocrine system.[20]

Another analyst of the history of the Odums' work, F.E. Golley, also confirms that at many times the metaphors in use by both Eugene and Howard Odum were mechanical and that this can also be said of most systems ecologists and ecosystem ecologists in the 1950s and 60s.[21] Ecosystem scientists, says Golley, were generally unafraid of combining mechanistic and organic analogies. Ecosystems were, for example, variously compared to the networks of electrical circuit boards, organismal and cellular physiological systems, and automatic machines.

Eugene Odum delivers these metaphors as much as anyone but sometimes he goes further than that, indicating that machines are not merely metaphorical when compared to ecosystems but in fact analogous at the level of fundamental laws. For example, he is quoted as saying that:

> The relationships between producer plants and consumer animals, between predator and prey, not to mention the numbers and kinds of organisms in a given environment, are all limited and controlled by the same basic laws which govern non-living systems, such as electric motors and automobiles.[22]

Odum, when he said things like this, might be interpreted as being in a historical point of ecology (the 1950s and 60s) that required a mechanical pretense just to insure the success of his field during the mechanistic predilections of science at the time. The ecosystem concept, from its inception in the 1930s to its popular acceptance in the 1960s, was viewed by its practitioners and advocates as a modernisation program. The movement that was systems ecology was presented as a modernisation movement also by the Odums, and the famous 1953 Odum textbook on ecology, *Fundamentals of Ecology*, has been seen as a manifesto for such modernisation. Golley, for instance states:

> In America the ecosystem concept appeared to be modern and up to date. It concerned systems, involved information theory, and used computers and modeling. In short, it was a machine theory applied to nature.[23]

However, this is not the whole story behind Eugene Odum's apparent mechanicism since he continued his references to mechanistic analogies all his life and he also continued to make comparisons between organic and mechanical systems. He not only did this for ecosystems but all systems that ecosystems might be connected to. In this regard human populations, like cities and nations, were hypothesised by Odum, to run like ecosystems. As Steverson notes

> Odum even lists cities as examples of heterotrophic ecosystems. One can isolate the input of energy into the system (e.g. the burning of fossil fuels to produce electricity), its transfer along various pathways composing the system (the transmission of electricity along power lines and into homes), the various uses to which energy is put by constituent organisms (watching television, heating homes, powering appliances, etc.) and the dissipation of energy from the system (loss of

> heat from poorly insulated homes, unused lights left on etc.)...Ecosystemically speaking there is no difference between New York City and the Everglades. They are both simply physical systems, analysable in terms of patterns of energy production, transfer, and loss.[24]

What this may say to those who wish for an ecologically-relevant organic view of nature is not to focus upon the Odumian version of the ecosystem in order to find it.

Organicism and the Cyberneticists

The Odums were not the originators or the developers of the mixed organic/mechanical metaphor; they had it handed to them by the cyberneticists. As Golley notes:

> Ecosystem ecology also was formed, in part, in the languages of engineering and economics and in the new subjects of cybernetics and information.[25]

Cybernetics, as has already been discussed in Chapter Three, owes much to Norbert Wiener's projects on the theory and operation of advanced weaponry. When Wiener retired from service at the RAND corporation, however, his determination to apply cybernetics to other situations meant that he continued to explore the parameters of his new cybernetic ideas. Wiener, like holist organic thinkers before him and systems theorists after him, would talk of homeostatic mechanisms in both living settings and non-living settings, in both natural and artificial situations and with both mechanical and organic metaphors.

Today the ambiguity between organicism and mechanicism is continued by modern day promoters of cybernetics. For instance Capra praises what Birch, Ferre, Cobb and Davies would regard as organic thinking and incessantly delights in talking about the need to develop an holistic ecological worldview. Capra himself goes on to state that the science of ecology is "the mapping out of pathways of energy and matter in various foodwebs."[26] This is to say that while he advocates holism and organic-type thought in ecology Capra actually suffers from mechanicism by choosing an ecology rooted to mechanical materialism.

We can see this, for example, when Capra himself admits cybernetics has a mechanical heritage. For instance, when he tells his readers how the cyberneticists explained their concept of self-regulation Capra says:

> To illustrate the same principles with a mechanical device for self-regulation, Wiener and his colleagues often used one of the earliest and simplest examples of feedback engineering, the centrifugal governor of a steam engine. It consists of a rotating spindle with two weights ('flyballs') attached to it in such a way that they move apart, driven by centrifugal force, when the speed of the rotation increases. The governor sits on top of the steam engine's cylinder, and the weights are connected with a piston, which cuts off the steam as they move apart. The pressure of the steam drives the engine, which drives a flywheel. The flywheel, in turn, drives the governor, and thus the loop of cause and effect is closed.[27]

As the historian Otto Mayr says of this connection between cybernetics and the steam engine:

> When Norbert Wiener in 1947 christened his new science of cybernetics, he was expressly paying tribute to what he considered the earliest cybernetic device, the word governor is derived via the Latin *gubernator* from the Greek steersmen.[28]

Capra generally tries to down play the mechanical heritage of systems thinking by claiming that 'the main characteristics of systems thinking...were pioneered by biologists...in the 1920s...who emphasised the view of living organisms as integrated wholes'. Capra astutely uses the term 'systems thinking' instead of 'organic thinking' because he is only too well aware of the contradiction in labeling systems as organic. This does not stop a whole lot of other people, whose ideas are very similar to Capra's, like Birch, Cobb, Goldsmith etc, from interchanging the words 'organic' and 'systems' quite freely.

The ability for systems to be identified with both organicism and mechanicism is a consequence of the closeness that all three terms have with unity. They can be interchanged quite freely in public parlance without seeming at all contradictory. For instance one can describe the 'machinery of the state', the 'state as being an organism' and the 'state as a system' without sounding contradictory.

Pre-Cybernetic and Pre-Odumist Mixing of Metaphors

Since machines and organisms (and mechanicism and organicism) are so intimately woven when considering the history and theory of cybernetics and systems science we might ponder whether there was a time before systems science and before cybernetics when a pure form of organicism was around. The superorganism theory of Clements might be one example but Clements was really trying to 'physiologise' ecology: "for Clements, ecology was nothing more than 'rational field physiology'."[29] And as Hagen also points out, other natural historians that were Clements' contemporaries were also given over to mixing their metaphors:

> in fact, the ideas of nature as superorganism and nature as machine were often used interchangeably...For Forbes, the lake was both a complex machine and an organism.[30]

Stephen Forbes, like Clements, Gleason, Tansley and Lindeman, is another reputed 'founding father of ecology' who sits uneasily within one or the other of the camps of organicism or mechanicism. Organicist unitarians might like to claim Forbes as one of theirs for he talked much about the balance of nature and the unity of ecological communities. Forbes, indeed may be claimed as the first ecologist (thirty or more years before Clements) to have postulated about the self-regulatory processes that kept populations in a balanced unity. In doing this, however, Forbes, himself was influenced strongly by the famous Nineteenth century social thinker Herbert Spencer, who himself, had both organicism and mechanicism embedded deep within his writings about both society and nature. This intermixing of the organic and the mechanical was evident, amongst other places, in Spencer's work on population equilibrium (which was the work that Forbes used to map out his own 'balance of nature' ideas). Although Spencer's name is synonymous with the 'organic view of society'--which flowed to and fro with regard to his similar thoughts on nature--Spencer, was originally trained as an engineer and maintained himself as 'an assiduous...inventor of gadgets'[31] who 'thought of nature as a moving equilibrium between opposing forces'[32] like those evident in Nineteenth century physics. If the metaphysics of equilibrium in Spencer's work is traceable to his machine heritage then we may assume that this had an affect on his own writings on balance as well as that of his readers like Forbes. Where Spencer used his equilibrium concepts to promulgate the idea that industrial society was working towards a harmonious balance like that already achieved in a homeostatic organism, Forbes posited that such self-regulation was also a tendency in natural communities. And just as Spencer uses concepts of 'systems' and 'mechanisms' which are derived from the mechanical sciences, so Forbes uses mechanical metaphors in his work on actual ecological communities.

As well as this it can also be claimed that when Spencer is being explicitly biological (and thus explicitly organic according to many New Scientists and Postmodern Scientists) his organicism is still linked with mechanical concepts since the biological metaphors he is using (to describe either social phenomena or biological phenomena) are in fact today regarded by environmental sympathisers as being mechanical. This is to say Spencer's emphasis on functions, his emphasis on physiology, his emphasis on systems, could only have come about if he prescribed to mechanical ideas of biology. Functions and systems, after all, are traceable to earlier 'mechanical' ideas that variously flowed from mechanical sciences such as astronomy, physics and engineering, or from mechanised biological disciplines such as physiology.[33]

Spencer's mixing of organicism and mechanicism in the late Nineteenth century are paralleled by Edward Goldsmith's social and natural philosophical ruminations in the late Twentieth century. Goldsmith, as we have seen, has berated traditional ecology for being mechanist whilst praising Odumian ecosystem ecology for being organic. When doing this he tries to characterise the organic/mechanic dichotomy with regards to pre-Odumist players in ecology, i.e.: before Odum ecologists were mechanicist. This dichotomisation, however, also fails. Goldsmith picks on the early Twentieth century mathematical biologist Alfred Lotka as being reductionist and mechanist, since Lotka's quantitative studies of population biology seem, to Goldsmith, to treat nature and its members in a robotic manner. However, Lotka operated within a metaphysical schema so similar to Goldsmith's that Goldsmith might end up criticizing himself. Like Goldsmith, Lotka believed evolution to be a universal biotic principle and thought "it would be profitable to analyse the evolution of the entire world system as a whole."[34] And like Goldsmith's heroes, Lovelock and Odum, Lotka "compare the world to a giant engine or, using an image familiar in thermodynamics, to a giant mill wheel."[35] And like Goldsmith (and Lovelock and Odum), Lotka also wanted humans to be partnered with nature, as collaborators rather than manipulators, to altruistically contribute to nature's efficient running of energy and materials.[36] And also like Goldsmith, Lotka wanted society not to float free from nature but to be grounded in the laws of nature.

What is also quite noteworthy is that Lotka got his insights not from ecology itself but from the mathematical economics of early Twentieth century capitalist economists. This may be interpreted as Lotka being unlike Goldsmith from a moral point of view but, as has been indicated in earlier chapters, very much like him from a theoretical and historical point of view since Goldsmith's metaphysics might be just as much a construction of capitalist/liberalist philosophy as it is a construction of contemporary organic ecology/environmentalism.

What is also of interest to us with regards to the topic of this chapter is that mechanicism, for Lotka, is not at odds with unity but very workable within unity since mechanicism, to him, is completely capable of working with ideas of teleology, interconnectedness and functionalism that we find so prevalent in organic unity thought. Lotka for instance saw the whole world as one vast machine and he spent much of his time searching for efficient energy flows within it (much the same sort of task that Lovelock performs today).

If Goldsmith decries Lotka as mechanist then how does another of Goldsmith's favourite founding fathers of modern ecology, Frederick Clements, stand up? Firstly, we should note that:

> Late in his career, Clements dabbled in philosophical holism, but his physiological perspective actually reflected an extreme form of mechanistic reductionism. At all levels--individuals, species, or community--Clements explained change in terms of simple, stimulus-response reactions.[37]

So it appears that Clements delved into both organic and mechanical science too, even as adjudged from Goldsmith's own criteria for distinguishing those two things.

Other important figures in ecology also exhibited what looks like, to modern day sensibilities, similar oscillations between mechanicism and organicism. For example; when describing the balance of nature (a very unitarian idea that Goldsmith has much sympathy for), the important mid-Twentieth century ecologist A.J. Nicholson chose to express it not in organic terms, as Goldsmith would see fit, but in mechanical terms:

> Nicholson believed that animal populations are normally in equilibrium and fluctuated only within restricted limits. He used a gas law analogy of a balloon to illustrate how a population would change if its environment changed but would rapidly become into equilibrium with the new environment, presumably as it stabilised.[38]

And we might also note once more that Arthur Tansley, the inventor of the ecosystem concept (which is viewed by so many unitarians as being intrinsically organic), actually dreamt up the concept in order to mechanise ecology against its organic sensibilities within Clementsianism. Although "Tansley's sophisticated mechanistic view of nature also retained strong organismal overtones."[39] Tansley often suggested that "all living organisms may be regarded as machines transforming energy from one form to another."[40]

As has been alluded, the muddled ambiguity between organicism and mechanicism is not just a thing of the past, it is alive and kicking today when scientists make philosophical statements. For instance, listen to the holist sympathiser, Rollo, as he explains from where modern ecology might gain better holistic principles:

> To forge a unified science of biology, a broad base of understanding that transcends scales is required. Biologists can learn some strong lessons from engineers.[41]

If unitarianism may express its holistic ideas via both mechanicism and organicism without fear of contradiction then what does that say about Gaia, which organicists claim to be the largest organic being in existence? If organicism is so affiliated with mechanicism so that we can hardly speak of one without recourse to concepts derived from or applicable to the other, and if James Lovelock's cybernetic heritage is just as imbued with these mixed metaphors, then Gaia is not just the ultimate living organism as Capra and Goldsmith would state; it is also the ultimate machine.[42]

That Lovelock might envisage Gaia in this very vein can be observed when he states:

> In this book I often describe the planetary ecosystem, Gaia as alive, because it behaves like a living organism to the extent that the temperature and chemical composition are actively kept constant in the face of perturbations. When I do I am well aware that the term itself is metaphorical and that the Earth is not alive in the same way as you and me, or even a bacterium. At the same time I insist that Gaia theory itself is proper science and no mere metaphor. My use of the term 'alive' is like that of an engineer who calls a mechanical system alive to distinguish its behaviour when switched on from that when switched off, or dead.[43]

Lovelock's mechanistic heritage is further betrayed--quite openly--when Lovelock himself describes his program in science as an extension of physiology to the biospheric level; physiology, of course, being a prime site of mechanism within the biological sciences. Just as William Harvey's mechanical renderings of human blood circulation announced the arrival of physiology to the scholarly world, so Lovelock wishes to extend and advance Harvey's insights to invent geophysiology; the physiology of the globe.[44]

Furthermore, Lovelock chooses to "describe Gaia as a control system for the Earth--a self-regulating system something like the familiar thermostat of a domestic iron or oven"[45] I suggest that it would be hard for any holist environmentalist or Postmodern Scientist to interpret irons or ovens in an organic framework.

Conclusion

There are numerous ways to look at the organicist/mechanicist dichotomy. The first is to say that organicism and mechanicism are completely separate and different ways of looking at the world. This is the way Postmodern Scientists look at the dichotomy. This may be the only wrong way of looking at the organic/mechanic dichotomy. However, there are several right ways. You can look at the dichotomy as cloudy and see that the organicism as presented in science is actually mechanical and then try and find the right (pure) type of organicism. Alternatively you can say (using mixed metaphors to highlight the implicit irony) that organicism is infected with mechanical bits and pieces that must be severed off and ejected. Another alternative is to break down the inherent objectivity of 'organicism' and 'mechanicism' altogether and then go on to say that they are virtually the same things. Under this view, and with regards to Twentieth century science, to call something organic is to invoke mechanical metaphors and to call something mechanical is to invoke organic metaphors. This is most stark in relation to the science of ecology and its attempts to form ideas about ecosystem structure and function. What defines both organicism and mechanicism through these metaphors can always be applied to the other.

What this deconstruction of organicism and mechanicism does is to unsettle the ecopolitical appeal of organicism. If New Paradigmers, New Scientists, and Postmodern Scientists like Capra, Birch, Ferre, Cobb, etc. are to appeal to the ethical righteousness and metaphysical rightness of organicism, then they also appeal to the normative righteousness and metaphysical rightness of mechanicism. The construction of a worldview that is supposed to be environmentally friendly must be aware of this mixing of philosophies and must be aware that any truly new environmentally-friendly scientific narrative might have to be both post-mechanist and post-organicist.

Notes for Chapter 7 (Organicism Vs Mechanicism)

1. See, for example, P. Davies, *The Cosmic Blueprint*, (Harmondsworth: Penguin, 1987); C. Birch, *On Purpose*, Sydney: UNSW Press, 1990); E. Goldsmith, *The Way: An Ecological Worldview* (NY: Shambhala, 1993); D. Zohar and I. Marshall, *The Quantum Society: Mind, Physics and a New Social Vision* (London: Flamingo, 1993); F. Ferre, *Hellfire and Lightning Rods: Liberating Science, Technology and Religion* (NY: Orbis Books, 1993); D. Bohm, "Postmodern Science and a Postmodern World", in C. Merchant, ed, *Key Concepts in Critical Theory: Ecology* (NJ: Humanities Press, 1994) pp.342-350; A. Gare, *Postmodernism and the Environmental Crisis* (London: Routledge, 1995), F. Capra, *The Web of Life*, NY: Harper and Row, 1996).
2. R. D. Moore, "From Science to Mythology: A New Vision of Reality", in J. Kliest and B.A. Butterfield, eds, *Mythology: From Ancient to Postmodern* (NY: Lang, 1992).

3. For example, see E.P. Odum, *Fundamentals of Ecology* (Philadelphia: Saunders, 1971); J. Lovelock, *Healing Gaia: Practical Medicine for the Planet* (NY: Harmony, 1991); Birch, *On Purpose*; Goldsmith, *The Way*; F. Ferre, *Hellfire and Lightning Rods*; A.J. Fabel, "Environmental Ethics and the Question of Cosmic Purpose", *Environmental Ethics*, 16, (1994) pp.303-314, R.U. Ayres, *Information, Entropy and Progress* (NY: AIP Press, 1994); U. Merry, *Coping With Uncertainty: Insights from the New Sciences of Chaos, Self-Organisation and Complexity* (Westport: Praeger, 1995); Capra, *The Web of Life*.
4. J.B. Cobb, "Ecology, Science and Religion: Toward a Postmodern Worldview", in D.R. Griffin, ed, *The Reenchantment of Science* (Albany: SUNY Press, 1988) p.122.
5. R.P. McIntosh, *The Background to Ecology*, (Cambridge: Cambridge University Press, 1985)
6. T. Benton, *Natural Relations: Ecology, Animal Rights and Social Justice* (London: Verso, 1993) p.14.
7. Bohm, "Postmodern Science and a Postmodern World", pp.343-344.
8. Birch, *On Purpose*, p.57.
9. Cobb, "Ecology, Science and Religion: Toward a Postmodern Worldview" p.107.
10. Capra, *The Web of Life*, p.37.
11. *ibid*, p.50.
12. Birch, *On Purpose* p.75.
13. D. Bohm, "Postmodern Science and a Postmodern World", in D.R. Griffin, ed, *The Reenchantment of Science* (Albany: SUNY Press, 1988) p.60.
14. And see McIntosh, *The Background to Ecology*; J. Hagen, *An Entangled Bank: The Origins of the Ecosystem Concept* (New Brunswick: Rutgers University Press, 1992); F.B. Golley, *A History of the Ecosystem Concept in Ecology* (New Haven: Yale University Press, 1993); D. Worster, *Nature's Economy: A History of Ecological Ideas* (San Francisco: Sierra Club, 1994).
15. Golley, *A History of the Ecosystem Concept in Ecology*, p.107.
16. Goldsmith, *The Way*, p.xix.
17. *ibid* , p.54.
18. *ibid*, p.191.
19. S. Rowe, "From Reductionism to Holism in Ecology and Deep Ecology", *The Ecologist* , 27, 4, (1997) :147-151.
20. Hagen, *An Entangled Bank*, p.135.
21. Golley, *A History of the Ecosystem Concept in Ecology*.
22. Odum, in Hagen, *An Entangled Bank*, p.136.
23. Golley, *A History of the Ecosystem Concept in Ecology*, p.2.
24. B.K. Steverson, "Ecocentrism and Ecological Modeling", *Environmental Ethics*, 16, (1994) p.86. Steverson also details how this approach by Odum is inimical to environmental evaluation:

> "if taken seriously and placed in the context of an ecosystem perspective, the ecocentric claim that moral consideration should be extended to our ecosystemic flows and to the ecosystem as a whole might entail extending

> moral consideration to such things as power stations and telephone lines. I assume that such a conclusion would be unacceptable to ecocentrists."

25. Golley, *A History of the Ecosystem Concept in Ecology*, p.67.
26. Capra, *The Web of Life*, p.172.
27. *ibid*, p.61.
28. O. Mayr, *Authority, Liberty and Automatic Machinery in Early Modern Europe* (Baltimore: Johns Hopkins University Press, 1986) p.195.
29. Hagen, *An Entangled Bank*, p.15.
30. *ibid*, p.9.
31. S. Andreski, *Introduction to: Principles of Sociology by Herbert Spencer* (London: Macmillan, 1969) p.ix-xxxvi.
32. S.E. Kingsland, *Modeling Nature* (Chicago: Chicago University Press, 1985).
33. See C.J. Glacken, *Traces on the Rhodian Shore: Nature and Culture in Western Thought From Ancient Times to the End of the 18th Century* (Berkeley: University of California Press, 1967); P. Mirowski, *More Heat Than Light: Economics as Social Physics, Physics as Nature's Economics* (Cambridge: Cambridge University Press, 1989); P. Christensen, "Hobbes and the Physiological Origins of Economic Science", *History of Political Economy*, 21, (1989) :689-709.
34. S.E. Kingsland, "Economics and Evolution: Alfred James Lotka and the Economy of Nature", in P. Mirowski, ed, *Natural Images in Economic Thought*, pp. 232.
35. *ibid*, p.234.
36. This point, and the following points, in reference to Lotka are drawn from Kingsland, "Economics and Evolution: Alfred James Lotka and the Economy of Nature".
37. Hagen, *An Entangled Bank*, p.23.
38. Macintosh, *The Background to Ecology*, p.188.
39. Hagen, *An Entangled Bank*, p.80.
40. Tansley, quoted in Golley, *A History of the Ecosystem Concept in Ecology*, p.218.
41. C.D. Rollo, *Phenotypes: Their Epigenetics, Ecology and Evolution* (London: Chapman and Hall, 1995).
42. This does not mean Capra would think it suitable to absolutely classify Gaia as the ultimate machine with no qualifications since he does, at times, make admissions as to the differences between machines and organisms. Witness this passage from Chapter 8 in *The Web of Life*:

> "The first obvious difference between machines and organisms is the fact that machines are constructed, whereas organisms grow. This fundamental difference means that the understanding organisms must be process-oriented. For example, it is impossible to convey an accurate picture of a cell by means of static drawings...Cells, like all living systems, have to be understood in terms of processes reflecting the system's dynamic organisation. Whereas the activities of a machine are determined by its structure, the relation is reversed in organisms--organic structure is determined by processes. Machines are constructed by assembling a well-defined number of parts in a precise and pre-

established way. Organisms, on the other hand, show a high degree of internal flexibility and plasticity. The shape of their components may vary within certain limits and no two organisms will have identical parts. Although the organism as a whole exhibits well-defined regularities and behaviour patterns, the relationships between its parts are not rigidly determined. As Weiss has shown with many impressive examples, the behaviour of the individual parts can, in fact, be so unique and irregular that it bears no sign of relevance to the order or the whole system. This order is achieved by co-ordinating activities that do not rigidly constrain the parts but leave room for variation and flexibility, and it is this flexibility that enables living organisms to adapt to new circumstances".

43. Lovelock, *Healing Gaia,* p.6.
44. *ibid*, p.10.
45. *ibid*, p.11

CHAPTER 8

An (Other) Postmodern Ecology

Introduction

If the unity of nature concept is a metanarrative which is unable to do the things that environmentalists ask of it, i.e.: value and protect the environment, then where may we look for one that does? Can we write an alternative narrative that does not rely on systems, physicalism, functionalism, anti-individualism, self-regulation or progressive evolution and which does not posit ontological unities, hierarchies or organic/mechanical balances? Is there an alternative that might jettison the metanarratives of Modernism, an alternative that we might properly call postmodern? This chapter recounts some of the criticism drawn in Chapter Six in an attempt to identify what such a narrative might have to look like in order to live up to contemporary postmodern thought.

The first thing that a postmodern environmental narrative might have to give up is its 'meta-ness'. Given that postmodernism is not at all fond of universalising tendencies, it is appropriate to restrict our metaphysical schemes just to a part of the universe. To this effect this chapter can be seen as addressing that part of the universe that unitarians are most fond of uniting in order to pursue ecopolitical endeavours: the terrestrial forest communities of the world. As we all know, the science that narrates about this part of the universe, and the science which, as we have seen, has been the playground for unitarian thinkers, is ecology. This chapter, then, might variously be thought of as

1) an attempt to document what a postmodern ecology might be responsive to and acknowledging of;

2) an attempt to postmodernise ecology,

3) an attempt to proffer an example of what a postmodern ecology might possibly look like.

When doing this it must be remembered that this will not be the only postmodern ecology, or postmodernisation of ecology, that may exist, since:

> An infinite combination of alternatives allow different and varying ways to put together the elements that constitute postmodernism.[1]

To start with we need to identify what a postmodern ecology would need to be appreciative of. Here we may draw on the analysis provided in Chapter Six to assert that a postmodern ecology would appreciate ephemerality and change, dissent and disunity, atypicality and strangeness, and non-universal story-telling. A postmodern ecology might also be thought to exhibit at least some of those one-word characteristics of postmodernism detailed in Chapter Six. For example, a postmodern ecology might be anti-scientific, anti-systems, anti-hierarchical, anti-teleological, anti-foundational and inclined towards fragmentation, heterogeneity, absence, aporia, pastiche, meaninglessness, while working toward a multidimensional exploration of Otherness.

To do this with any effect (and as alluded to throughout Chapters Two to Seven) a postmodern ecology would also be non-mechanist/non-organic and non-functionalist whilst being appreciative of chaos (without supposing such chaos gives rise to order). Within such a postmodern ecology, each pattern, operation, behaviour, process, object or subject would have to be explained in terms of its own local narrative history rather than as contributions to an overarching unity, mega-purpose, or meta-process. Such a postmodern ecology would also be aware of its own metaphoricity; its own social creation as an open and malleable tool in ecopolitical rhetoric and action. There may be a tradition within ecology that does all this, or at least has the capacity to do so; Gleasonian ecology.

In this chapter a postmodernised Gleasonian ecology is drafted out. Such a postmodernised Gleasonian ecology is not presented as offering a superior project for the philosophy of nature but as an alternative that does not rely on unity, one that utilises atomism in a positive environmental way (something which unity intellectuals have maintained couldn't be done).

A Postmodern Gleason?

It might be recalled, from earlier chapters, that Gleason's ecological writings were a reaction to unity and uniformity in professional ecological theory. In contrast to Frederic Clements and other superorganicists, Gleason maintained that there is not a lot of order in ecological communities and hardly a direction of change towards a perfect climax. This seems to exhibit some resonance to comments by some postmodern intellectuals. For instance, when Zygmunt Bauman makes the following statements, which we have already come across in Chapter Six, he could be talking about Gleason's ecology just as much as about the changes in society he has witnessed in the late Twentieth century:

> 'Postmodernity' being a semantically negative notion, defined entirely by absences--by the disappearance of something which was there before--the evanescence of synchronic and diachronic order as well as of directionality of change.[2]

The Gleasonian appreciation of disunity accompanied by its pervasive incredulity towards large structures might potentially lead us into investigating Gleason's ecology as a postmodern form of ecology. Viewing the terrestrial assemblages of plants and animals in the world as anarchic patches of unrelated individuals (and groups of individuals) does confer upon Gleasonianism a certain resonance with postmodern art and its penchant for pluralistic pastiche and muddled mosaic. We could also note that Gleason's recognition of the typicality of non-conformity--rather than the subjugation of non-conformity into the realms of negligibility and exception--also, at least superficially, resembles postmodern social and political theory.

If "the very foundation of postmodernity consists of viewing the world as a plurality of heterogeneous spaces and temporalities"[3] and if postmodernism is held to seek "the uniqueness of the parts rather than the unity of a theoretical whole"[4] then it is probably legitimate to acknowledge the postmodern credentials of Gleasonian ecology over those of Clementsian ecology.

However, such a fracturing of theoretical unity does not necessarily contribute to the other major postmodern aim: the fracturing of metanarratives. Just because there is a dissolution of unity within Gleasonianism does not mean that there is a dissolution of foundational ideas. Although fragmentation lends itself to the extirpation of foundational principles[5], a fragmented view of the world's ecology does not, by itself, sufficiently derail foundational ecological theory. If we extend Gleason's skepticism towards superorganismic unity in ecology through a postmodern deconstructionist type of critique we may, however, dissolve the need for both ecological unity and a whole gamut of interrelated foundational principles that prop it up and support it. When doing this we introduce, in this particular case, what might be named 'postmodern associationism'.[6]

According to Tester the "dismantling of bounds and boundaries is an inherent project of postmodernity."[7] While Gleason makes inroads toward breaking the ontological boundedness of ecological communities (by denying the fact that communities are bounded), a postmodern critique would extend the breaking up of boundaries to the epistemological realm. In modern ecology, epistemological boundedness is most explicit in the idea of 'levels of biological organisation'. As we have seen the orthodox way to present this is through a hierarchy framework like this:

cell--organ--organism--population--community--ecosystem--biome--biosphere.

Under unitarianism the varying levels of biological organisation are essentially the same phenomenon. Individual organisms are much the same as ecosystems, they merely operate, or are characterised, at different scales. This is evident in the systems theory of both Lovelock's Gaian ecology and also in the Odums' ecosystems studies. For example, we can spot this as a working metaphysical viewpoint in Eugene Odum when we note Hagen's comments about Odum's doctoral training. Hagen quotes Eugene Odum as saying:

> The transition from bird physiology to ecosystem function was quite natural for me since it involved moving up the hierarchy from physiological ecology of populations to physiological ecology of ecosystems. It's really not such a big step to go from whole organism metabolism to community metabolism.[8]

Here we see how Eugene Odum's particular conceptualisation of 'levels of organisation' follow fairly strictly the framework outlined just above. An ecosystem is virtually the same as an individual bird. All the various organisms of an ecosystem become unified into an individuated unified entity that is Odum's ecosystem.

Under Gleasonianism, however, any properties of individuation (such as self-preservation, teleological behaviour, autopoesis, psycho-physical unity *etc*) afforded to the levels of organisation above the individual can only be maintained with much analogical maneuvering. Due to the indefinite and unbounded character of Gleasonian ecological communities, the very concept of hierarchically-arranged levels of biological organisation can be brought in to doubt. While levels may be heuristically convenient and pedagogically useful they also compartmentalise and unify spatio-temporal phenomena that are much more heterogeneous and unbounded than the concept of the level signifies.

In many breeds of ecology (most notably community ecology) the concept of the level is acknowledged as a mere abstraction, flowing, as it does, around, over, and under the many nuances of ecology to which it cannot apply. Unitarians, however, tend to use the concept in a totalizing manner, envisaging biological levels to exist as clearly defined layers like floors of a modern building. The various levels are held to be variously connected by ladders of information[9], staircases of feedback mechanisms[10] or elevators of energy.[11] This view of the world's ecology is one of a well-ordered and intricately organised hierarchy whereby the process of downward causation acts to keep all lower levels in line:

> as smaller units integrate and aggregate into larger units, so they give rise to new rules, which in turn constrain and regulate the component subsystems to comply with collective behaviour of the system as a whole.[12]

Goldsmith expresses the idea of downward causation like this:

> larger systems, from Gaia downwards, control and co-ordinate the behaviour of their constituent parts...[and the]...adaptive strategies [of these parts] are all geared to maintaining the stability or homeostasis of the entire Gaian hierarchy.[13]

Capra, in turn, would say that order at one system level is the consequence of self-organisation at a larger level.[14]

A postmodern alternative would want to obliterate such an orderly and hierarchical conceptualisation of ecology. The postmodern penchant for fragmentation, heterogeneity and anti-hierarchy would also reject any attempt to place individuals and groups of individuals within neatly defined hierarchical levels.[15] If explicit metaphor is required, rather than envisaging the organisation of the biotic world as a well structured building, postmodern ecologists might choose to follow a path of de(con)struction and disintegration so as to view the ecological world as resembling a broken and fragmented jumble of variously shaped and precariously perched shards of debris in whose nooks and crannies the lives of individuals are contextualised and from whose micro-fissures reflections of localised worlds are ricocheted.[16]

Many contemporary philosophers of nature like to invoke levels organised into a hierarchical scheme since it allows them to discern level specific laws; laws that operate at one level and not others. Given this the heinous crime of reductionism cannot be conferred upon them since it is willingly acknowledged that lower level laws (such as genetic rules or atomic behaviour) do not determine the operation of higher level laws (such as populations or ecosystems).[17]

Although such a hierarchical view of ecology might render reductionism a benign force, hierarchical conceptions tend towards breeds of essentialism and holism within ecological science which are every bit as totalising as reductionism.[18]

When utilising the epistemological framework of 'levels' to describe biological organisation, the member entities of levels tend to become described/characterised by the essential properties held to exist for their particular level and they are also deemed to operate according to level-specific laws that confer upon them their totality of identity. A population of beetles becomes essentialised to the characteristics that populations of any type of organism are said to possess. A jungle ecosystem is made to suffer the indignities of being considered fundamentally the same as any other ecosystem upon the planet. The brain cells of a bandicoot are held to be basically comparable to algal sex cells. Intra-level generalisation is held to exist over and above any intra-level differences.

This type of intra-level generalisation is exhibited in the Wu and Louck's paper that attempts to generalise modern day ecological theory. In this paper they state:

> Such a hierarchically organised system can be seen as a system in which levels corresponding with progressively slower behaviour are at the top, while those reflecting successively faster behaviour are lower in the hierarchy. Higher levels impose constraints on the lower levels, and thus can be expressed as constraints. On the other hand, the dynamics of the lower levels can be so fast that their signals are smoothed out at higher levels, and often can be treated as averages.[19]

What should worry us here is the keenness to average out the actions of lower levels. Despite the chaotic abundance of a multiplicity of organisms producing a multiplicity of emergent properties, these properties themselves are capable of producing a 'smooth' base upon which the actions of higher levels can rightly be assumed to be based.

A postmodern association view would not honour such intra-level generalisation and would argue that the purported levels have no fundamental laws or essential characteristics. Nor do they give rise to cohesive 'smoothable' emergent properties as Wu and Loucks have suggested. Thus, different ecological communities are more different from each other than they are similar. Similarly, ecosystems cannot be generalised into a few laws, and there is no one typical cell that all other cells aspire to.

If there is not a definable set of cohesive general features that can be invoked to characterise a particular level, then the actual existence of levels must be called into question. Each plant community, each marine ecosystem, each bird population must exist as a unique and non-essentialisable collection of individuals that must demand to be studied on their own terms. They cannot be bounded by the concept of 'level' without receiving decrees of generalisation which are ill-fitting. Thus the utilisation of the idea of levels of organisation (like populations, communities and ecosystems) must become vague and necessarily fluid.

Ungluing The Ecological World

The postmodern deconstruction of levels of organisation in nature is one way of countering the modern scientific preoccupation with order and unity. Another way is to identify and eject two other metanarratives of contemporary ecology; competition and co-operation, which are both very important to unity.

According to most spokespeople in the natural sciences there are two different ways of looking at the interactions that prevail in nature. The first way is to perceive nature from the Darwinian perspective; a competitive struggle for existence. The second may be described as the Kropotkinesque view[20], whereby the dominant interactive process in nature is cooperation or symbiosis. While the cooperative view of nature is a common line of thinking to many unitarians, Darwinism is still, by far, the dominant paradigm in ecological science when it comes to explaining inter-organismal relationships and the structure of ecological collections. Of course; Darwin said a lot more about evolutionary ecology beyond competition, and an exclusive emphasis on competition as being the only contributor to community structure is held to be unwarranted by most ecologists. Whilst the American biologist Jared Diamond has been cited as indicating that competition is virtually the only determining factor of community structure[21], other biologists feel that competition is only one factor, and not necessarily the most important one, in determining the structure of particular ecological communities. For instance; predation and herbivory[22], disturbance[23], parasitism[24], and symbiosis[25] have all been noted as being important factors in influencing community structure.

This last named process, symbiosis, is often hailed by Gaians and unitarian environmentalists as being the prime relationship in nature, such that it is said to be a major determinant of the biotic structure of ecological collections. Darwinians do not necessarily negate the importance of symbiosis in ecology but they do try to bring it into the fold of Darwinism by claiming that symbiotic relationships are naturally selected adaptations that give the symbionts a competitive edge over other organisms--allowing better rates of survival in the face of nutrient shortage, predation or environmental stress. Scientifically trained Gaians, alike, do not want to rid ecology entirely of Darwinism; they merely like to see natural selection as operating at higher levels than Darwinians have hitherto appreciated. Ecosystems, and even the whole biosphere, thus become the unit of natural selection, hinting at the immense co-operation that must go on to achieve this.

Whether competition or cooperation is seen as the underlying expressive force in nature, both Darwinism and Gaianism can develop causal explanations about nature from these metatheoretical bases. Cooperation and competition are needed to glue the individuals to the whole. Their action in nature is to work to supply a uniform set of emergent properties which give rise to the community or ecosystemic level of organisation.[26] Getting rid of co-operation and competition will undo the unity concept in ecology by eliminating the need for intense (competitive or co-operative) relationships between all nature's members and by breaking up the levels of biological organisation whose very existence relies on the identifiable emergent properties that cooperation and competition are said to produce.

Yet despite the supposed omni-presence of competition and cooperation in nature, it could be argued that one of the most common relationships in ecological settings is non-interaction. Yes, a particular predator may hunt and kill a particular prey, and yes, a particular tree may compete with another particular tree for light, and yes, a particular alga might live symbiotically with a particular fungus. But it is just as possible to say that the most prevalent relationship in nature is no relationship. There are no connections between most organisms in the world. They exist in their own environmental context of course, whereby they interact on a local and capillary scale, sometimes in strangely-contorted and complicated networks, but any organism is profoundly disconnected with nearly all the other organisms in the world. Networks are never complete, never total. If they exist, they exist in a state of extravagant aporia.

While this book suggests ecological interactions are much more variable than the physicalism of systems ecology and Gaianism presuppose, the prevalence of fragmentation and disparity between the world's ecological members must be emphasised so that interactions and interconnections are, many times, non-existent. The absence of interaction, however, does not make the world a simpler place. It makes it more chaotic (latin reminder: the *cha* in chaos denotes absence) since there is a constant fragmentation of cause and effect in a world thought to be intimately linked.

The connectionism so apparent within ecological thought has been built up to convey how the delicate web of nature is prone to collapse if ill-inspired tampering with nature is undertaken by human beings. The postmodern view that is presented here, however, does not see the environmental crisis as the result of misguided tampering with the delicate balance and interconnections evident in nature but as a wholesale destruction of the ecological world itself. The forests of the world are not collapsing due to the web destroying nature of pollutants and micro-disasters which destroy a particular ecosystem by untangling the various networks that make it up, they are being destroyed by a sweeping and blanketing destruction that eliminates all members of an ecological community in near unison. Forests are mowed down in bulldozer-like fashion, rather than slowly advancing towards a state of moribund morbidity due to the surgical removal of species.

What this might suggest is that there is not one great big global ecological disaster lying over the horizon in the future due to ongoing disentanglement of ecological networks (as Lovelock would suggest) but instead there are bundles of smaller concentrated disasters, each itself a travesty, occurring right now. This is to say that extinction and environmental disaster, if we are to regard them as ethical issues at all, are local phenomenon. The environmental crisis occurs every day, when a beetle becomes extinct, when a forest gets swamped by a dam, when a lake gets polluted by industrial waste: these events are the environmental crisis in action, here and now. In contrast, Lovelock would have us believe that the environmental crisis is only a potential event; something that will happen one day if we keep doing these small-scale things so that altogether they cause a change in the Earth's global environmental conditions. Such a 'big picture' view of the environmental crisis as forever imminent (yet always abstract) rather than actually happening devalues the daily tragedies that afflict the world's ecological settings.

The Postmodern Ecological Individual

The destruction of levels, as well as competition and cooperation, in nature releases individuals and their aggregates from being trapped in an over-lording holism whereby their lives are constituted according to a greater whole. Downward causation cannot exist in postmodern ecology since there are no higher levels from which to be dictated to or determined by, only a heterogeneous cascade of shard-like collective features whose effects are ungeneralisable, lawless and self-contradictory and whose interactions with individual organisms and their environments are fragmented in scale and intensity.

If individual organisms are thus freed from the action of downward causation they are dissociated from the actions of any greater ecological whole. Such wholes only exist in the mental and social milieu of humans. As some postmodernists have alluded, postmodernism sometimes has the tendency to shatter structural entities to near non-existence[27] whereby the only important agents of any theoretical or normative relevance are individuals. This emphasis upon ecological individualism tends to concur with other postmodern analysts who maintain that while the cohesive identity of the autonomous Modernist subject may be pronounced dead, "the postmodern individual is still very much alive"[28] so that postmodernity has become the 'Golden Age of the Individual.'[29]

However, such fragmentation towards individualism could be seen to represent an alternative foundationalism by appealing to individualism as absolute reality.[30] This appeal would give heart to many a neo-Darwinist who would see in such a postmodern deconstruction of nature, philosophical support for natural selection (which is founded upon the sanctity of the individual as being the basis of the selection process). Darwinian evolutionary biology also posits that the unitary nature of the individual is manifested through its genetic essence. An individual equals this genetic essence and vice-versa.

On a closer inspection than that which Darwinists are willing, however, it becomes apparent that the genetics of an organism does not solely define an individual's life. Nor does the 'selfish gene' even lay down the parameters of an organism's existence. An organism is much more pluralistic than is generally accepted. If the genetic program of an individual does not enable it to feed from a particular energy source, the organism is not about to give up the ghost and submit to its pre-ordained genetically prescribed extinction. It goes on attempting to escape the essence reputedly embodied in its genes. It tries other food sources in order to do what it wants...survive, propagate, lie around in the sun...whatever. It has no essence, no essential drive to survive or to reproduce. It is essence-less and has a myriad of other things that it might like to get up to.

By acknowledging the relevance of a certain degree of ecological atomism in the description of ecological communities it might also be assumed that postmodern associationism tends to give rise to a coherent and cohesive notion of an ecological self; an organism is a rational and autonomous actor blessed with a high degree of ecological agency and a fixed identity. However, postmodern associationism might choose to parasitise upon the postmodern predilection towards schizophrenia to assert that ecological actors operate under a 'plurality of self'. The Gleasonian view of natural history might be interpreted to confirm that in any one ecological situation an individual can act in one (or more) of a multitude of ways. The exact way which is enacted being just as influenced by extra-genetic factors as by the genetic boundaries held to exist within the organism.

We are all familiar with the extragenetic cultural activities of primates that might be cited to indicate the plurality of the ecological self; tool-using chimpanzees[31] and potato-washing macaques[32] for example, yet there are many other animal types that exhibit a plurality of self. From the milk-cap ravaging blue-tit[33] through the jumping spider that chooses different predatory tactics for the various prey it encounters[34] to the rock-leaping Gobi fish that jumps from rock-pool to rock-pool at low tide based upon topographical experience gained at high-tide[35], the pervasivity of extra-genetic ecological activities is quite evident. All this lends credence to the conclusion that: "It is also becoming increasingly clear that animals share with us potential for cultural adaptation"[36]

The plurality of self in plants is also demonstrable in the way that individual shrubs and trees subtly respond to the variations in spatial patterns, herbivory episodes and dispersal agent availability that they experience. These environmental variations cannot be predicted by genetic essences, thus implicating a great degree of sensitivity on the part of the plants to their environmental context. There is also a case to be made about the intricate chemical communications going on between plants which might bolster such plurality of self, as for instance, when plants talk to each other about the diseases that are strickening them so that other plants may somehow fend against the same problem.[37]

Microbes, too, exhibit a plurality of self where they are called upon to act according to extragenetic factors, for instance the irregularity of algal cell movements[38] and protozoan disturbance responses[39] might indicate a plurality of options afforded to members of these microbial groups. What this says, at least to those willing to accept postmodern associationism, is that organismal behaviour is an experiential rather than genetic process and that "at the moment one should accept with extreme caution any statement on the inheritance of behavioural traits" via instinct.[40]

It also says that far from being automatons with a deterministic genetic essence; non-human individuals somehow manufacture their own behaviours from their daily encounters and so exist like Rosenau's postmodern human individuals, submitting "to a multitude of incompatible juxtaposed logics, all in a perpetual movement without possibility of permanent resolution or reconciliation".[41]

Whether or not it is wise to assume that non-human organisms are manifestly capable of multiple self-construction that postmodernists believe of humans, it is at least worthy to consider them as capable of experiencing pluralised lives whereby they operate as though unbounded by any coherent anthropogenic laws. With the genetic blueprint gone, with the essences of instinct downgraded in importance, non-humans must attract the same sort of detailed pluralistic study as human individuals. Instead of existing as manifestations of their genetic heritage, non-humans become (as Hollinger says of postmodern humans):

> multiple, not fixed...with no overall blue print. The various multiplicities that constitute the self at a given time are marked in play and dance with each other.[42]

The Metaphors in a Postmodernised Gleasonian Ecology

Because of its emphasis on an individualistic explanation of ecological communities, and because of its penchant for atomistic deconstruction of larger ecological abstractions, the ecology of Gleason has often been held to be both mechanistic and reductionistic by holist ecologists. Goldsmith, for instance, has attacked modern day versions of Gleasonianism in this vein.[43] However, Gleason's associations are hardly mechanistic. The Gleasonian units (i.e.: individuals) of an ecological community are not hollow mechanical shells and they do not just bounce off of each other like solid particles. Nor would a postmodern Gleasonian insist upon chaotic or atomistic individualism as an absolute. Local unities are often constructed (and dissolved) by the peculiar activities of individual members and the groups they comprise.

What this tends to indicate is that postmodern associationism is not an ecological form of a billiard ball universe; comprised of crashing and clashing atomistic individuals operating according to predetermined laws.[44] Postmodern associationism does not "see the world as consisting of mindless, meaningless, totally determined physical bits and pieces that are non-purposive" as Charles Birch would claim of non-organic philosophies of nature.[45] Nor does postmodern associationism necessarily advance an ethical theory promoting the sanctity of the individual over an ecologically or socially constructed whole (but a postmodern ecology such as postmodern associationism would place emphasis on respecting the individual Other as arbiter of its own reality without imposing metaphysical imperialism under the guise of organic unity).

If absence and decentering are postmodern traits, as explored in earlier chapters, then it might also be noted how, in the postmodern Gleasonian ecological world, there is a distinct absence of centralised control. According to two community ecologists, Whittaker and Woodwell, who followed a neo-Gleasonian line in the 1960s:

> There is in the community no centre of control and organisation, and no evolution toward a control system.[46]

Whittaker and Woodwell were reacting to the rise of systems ecology in the 1960s and they were expressly interested in denying that there was a tendency for the biotic components of ecosystems to work with the abiotic components in order to evolve into a stable, self-controlling and self-maintaining system.

Systems ecologists may claim the same for the ecosystem but the point that Whittaker and Woodwell would make, and which prompted them to make the above statement, is that an ecosystem conception of nature is a centralised form of natural history since everything of importance is run via the central process of negative feedback.

The decentralisation within Gleasonianism might be made more visible if we attempt to graphically compare the worldview of unitarian ecology as presented by someone like Fritjof Capra and with the postmodern ecology described in this chapter. Capra's model would look something like figure 1 on the page 228.

In Capra's model organisms A, B, C, D and E contribute to emergent properties which are so hard and fast and unified that they form (or 'self-organise') into a coherent set of emergent properties that equals a unified entity (which may be Clements' community, Odums' ecosystem or Lovelock's Gaia). The unified entity then acts homogeneously (though homeostatic feedback) mechanisms to envelop, control and limit organisms A, B, C, D and E. All of this together then rolls progressively forward via evolution to ever more stability or higher complexity or more diversity.[47]

If we attempt the same sort of modeling with Postmodern Gleasonianism then we end up with something quite different (see figure 2 on the next page). In this alternative model, organisms A, B, C, D and E give rise to multiple, divergent and contradictory emergent properties which are not unified enough to act homogeneously in order to effect a controlling influence via uniform negative feedback on the organisms. This disorder is set amongst, and gives rise to, various 'directions'; increasing complexity, decreasing complexity, increasing stability, decreasing stability, increasing diversity, decreasing diversity, etc.

The Reflection Section

A single cohesive alternative to any reputedly debunked metanarrative sounds suspicious, especially to the postmodern mind. Some might suspect that in proposing postmodern associationism at the expense of the unity of nature idea, this book is merely replacing one totalising foundation for another. This concern is legitimate and there are a number of ways that it may be addressed. Firstly, it might be wise to restate the metaphorical nature of all human endeavours; our truths, our models, our maps and our speeches (in the vein of Lakoff and Johnson and Miller.[48] If this is so then we can no more regard postmodern associationism as a literal translation of the reality of nature than we can regard superorganicism, the New Sciences or Postmodern Science as sciences with privileged abilities to know and represent nature. If "a worthwhile application of metaphor...must begin with the awareness of its own problematic; its own metaphoricity"[49] then we must conduct explorations into postmodern associationism in the light of its own metaphorical nature. This thought will encourage (or condemn) a postmodern associationist towards modesty with regards to any authentic correspondence with nature that might be suggested. Or as Miller puts it:

> the rejection of a metaphor [like the unity of nature] is not the work of objectivity, rationality, science, let alone 'purer' language; it can only be the challenge and replacement of another metaphor.[50]

This said, it is still appropriate to advertise postmodern associationism as a worthwhile alternative metaphor that avoids the pitfalls of unitarianism.

On the issue of being yet another totalising foundation it may be asserted that postmodern associationism enables, or accepts, that each biotic member on the planet constructs its own version of reality (although whether that member is doing

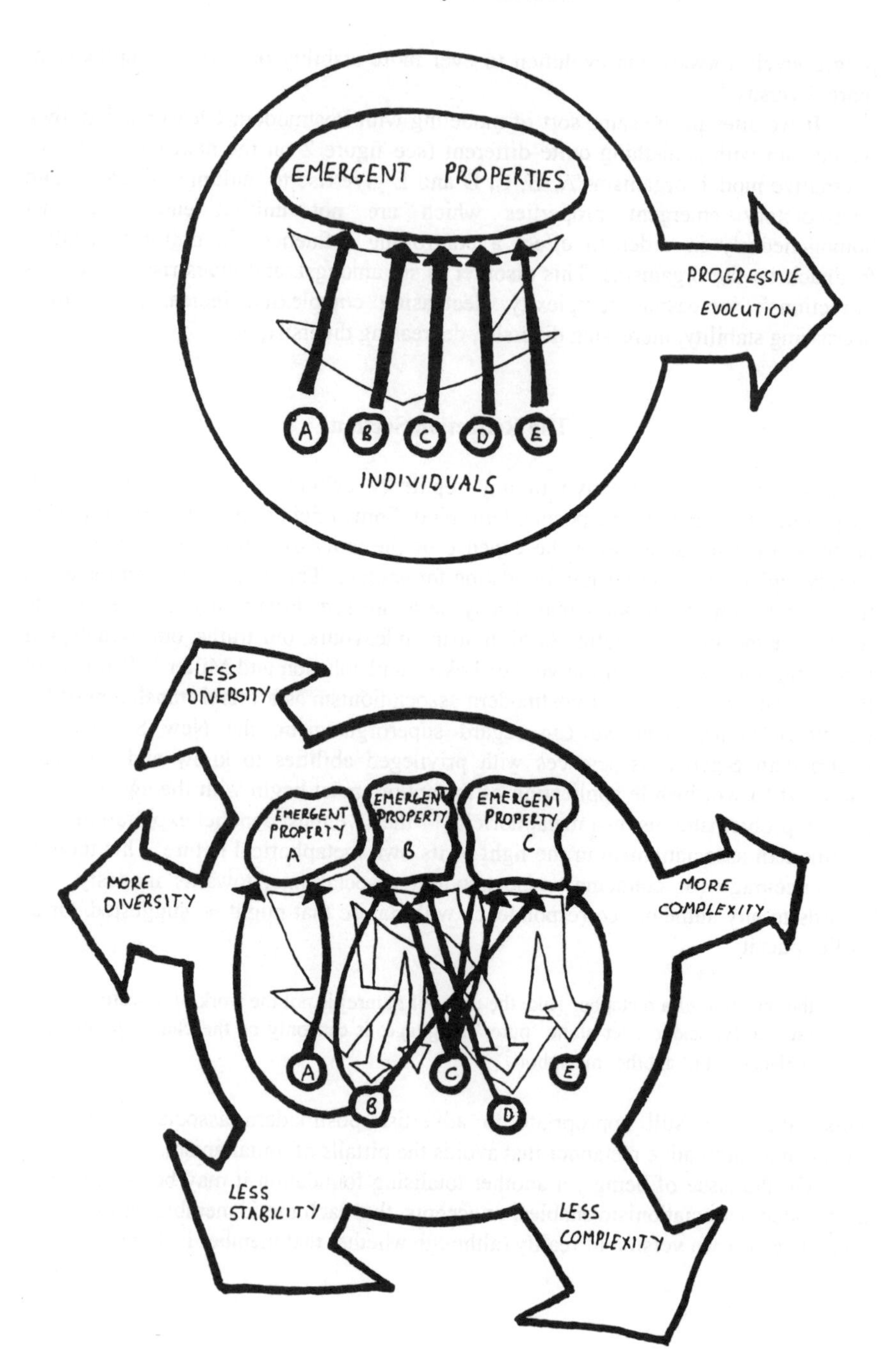
EMERGENT PROPERTIES
PROGRESSIVE EVOLUTION
A
B
C
D
E
INDIVIDUALS
LESS DIVERSITY
EMERGENT PROPERTY A
EMERGENT PROPERTY B
EMERGENT PROPERTY C
MORE DIVERSITY
MORE COMPLEXITY
A
B
C
D
E
LESS STABILITY
LESS COMPLEXITY

anywhere near all of the constructing is seriously open to debate) rather than having its reality thrust upon it purely by its functional position in a unity.
Postmodern associationism thus reflects a pluralistic way of looking at the world. It argues that we should not look to extra-galactic astronomy, quantum physics or science fiction to detect other universes. The plurality inherent on planet Earth reflects that every organism has its own de-centred, essence-less and fragmented reality. Each individual is its own universe with its own peculiar set of laws and generalisations. This suggests that we might need to redefine this thing that we live in as a pluriverse.

Furthermore, to counter claims that postmodern association is totalising it might be declared that the postmodern association outlook is not a worldview of the universe, nor a metanarrative of life or of all things living but merely a micronarrative story of some of the vegetation stands of the planet Earth. Thus it is a story that is only locally applicable from a spatial and disciplinary point of view, having very little to say that might be influential beyond the various sub-disciplines of the scientific tradition known as ecology and having nothing whatsoever to say about the character of the rest of the physical universe. Unlike systems science and unitarianism it makes no colonising forays into adjacent sciences and it certainly does not entertain the thought of being another 'New Paradigm' which must be adopted by all the world's human members in every single intellectual endeavour in order for an environmentally and socially benevolent society to be enforced.[51]

Students of the postmodern association outlook might also be unwilling to propose postmodern associationism as the only viable narrative of the world's natural history. Concessions are willingly made with regard to the relevance of ecosystem ecology and unitarianism by admitting the importance of emergent properties, ecological interaction and symbiosis in the practice of ecological study. And while Gleasonianism might like to see forests and other natural collections as heterogeneous mosaics varying unpredictably in composition throughout space and time, there is plenty of room within Gleasonianism for acknowledging the existence of uniform patches of nature within a particular area of ecological interest. (However, Gleasonian inspired ecologists would probably be more inclined to view these uniformities in composition as attributable to the adaptive significance and physiological tenacity developed by particular species rather than invoking holistic homeostasisas a cause). In addition the science of community ecology is well endowed with unity-like concepts (symbiosis, patches, guilds, coevolution) that emphasise the many close relationships within ecological collections--thus postmodern associationism may be interpreted as a molecular rather than atomistic view of ecology. We might compare such molecular natural history to Deleuze and Guattari's postmodern molecular epistemology, whereby they "advocate a molecular (as opposed to both an atomistic and macro) analysis."[52] This is not to deny that in the case of ecology some of these molecules can be amazingly long.

By constantly attacking the ecosystem outlook of nature this book might, perhaps, be thought a little cruel as it attempts to exterminate the totalitarian ecological principles associated with systems science. However, maybe it is possible to construct an alternative from the remains of such an extermination. Perhaps we should not try to wipe ecosystems ecology from the face of the Earth but merely downgrade its universality.

Using such an approach, and from the point of view of postmodernity's schizoid individualism (where the death of the Subject as a concretely conscious and stable identity is acknowledged but the multi-dimensional character of the Self is regarded as alive and kicking) it is easy for postmodern associationism to acknowledge that the ecosystemic dimension of the ecological self may, actually, be a transiently important part in describing the life and times of an ecological individual (whilst remembering that it is repugnant to postmodern sensibilities to regard the ecosystemic part as the *only* part worthy of consideration in ecological study or in environmental valuation).

There is also a need to point out that although there are many calls for community ecology and systems ecology to become more unified so as to elicit a more balanced scientific approach[53] this book is not one such call. Such calls allude to the possibility of a single intellectual utopia emerging that balances between the respective methods/metaphysics of community ecology and systems ecology which, when articulated, will supply humanity with an undistorted version of nature. All versions of nature, whether carefully crafted hybrids or not, must admit to being distortions through specific social/intellectual lenses. A systems-community ecology hybrid is just as loosely bound to the referent about which it narrates as either systems ecology or community ecology by themselves.

On the issue of foundationalism, and the postmodern desire to transcend it, we may adopt the sentiments of Patricia Waugh's statement about how many postmodern arguments are infected with alternative foundations and then go on to admit that postmodern associationism is just one such infected beast. Postmodern associationism might reject unitarianism, natural selection, levels of biological organisation, balance of nature.etc. as unifying principles in ecology but it still appeals too much to founding principles (methodological individualism, ontological 'being'-ness, or biocentrism, perhaps) to claim total ejection of foundationalism.

Alternatively we may wish to go beyond Waugh's statement and assert that every postmodern theory necessarily has its lurking foundationalisms since, despite claims to the contrary, any postmodern view that constitutes itself into a theory will appeal to some or other foundationalism, either because: 1) it wishes to survive in the academic world, 2) any theory has to make some suppositions about the nature of reality or knowledge, 3) theory is inherently unable to avoid constituting and defining that which it is explicating without pre-existing foundation-laden theory or 4) theorists are unable to escape the myriad of foundations contextualised into their lives.

And if postmodern associationism, or any other narrative of the world, is able to escape such problems of foundational transcendence it may only do so by miring itself into a foundationalism of anti-foundationalism. This is to restate a persistent problem within postmodernism that a belief in no foundations might be classed as such a strong and ever-present assumption that it is a kind of foundation in itself. (We might also argue, however, that the idea that we can never transcend foundationalism in narrative construction is itself such an absolute statement that it displays inherent totalisation).

These musings about postmodern associationism's desire to extinguish ecological foundations once again recounts Bauman's allusions to the ongoing existence of dilemma in postmodernity:

> behind the postmodern...hides a genuine practical dilemma: acting on one's moral convictions is naturally pregnant with the desire to win for such convictions an ever more universal acceptance; but every attempt to do just that smacks of the already discredited bid for domination.[54]

As explained before, where Modernism gets over its dilemmas by positing that all dilemmas lead to a better order, postmodernism accepts the insurmountability of these dilemmas without trying to sweep them under the mat of self-regulating unity.

Dissolving Humanity and Nature

Every political movement operating under Modernism has had a great and evil monster that it is out to slay. For socialists it is the capitalist monster, for feminists it is the patriarchy bogey-man, for black-activists it is the great goblin of racism. Environmentalists, too, have often embarked upon a battle to the death with an evil monstrous being; one called anthropocentrism.

This great and evil monster operates to undermine the true unity that the species 'humanity' has with the rest of nature, thus ensuring our anti-environmental attitudes and actions which have plunged the world towards destruction. This section aims to make a specific comment about this monster called anthropocentrism. To do this, an examination of the Humanity/Nature relationship is made.

The view of nature and humanity as promoted in this book is one of pluralism. Humanity/Nature is not dualistic as orthodox Marxists and capitalists perceive (whereby individuals may belong to either human existence or to the natural world).

Humanity/Nature is also not monistic as unitarians believe (whereby there is only one reality--that of unity--from which you can not escape). Humanity-Nature is actually pluralistic, whereby there are as many realities as there are members of the world; maybe more, given the plurality of self.

This postmodern emphasis on a plurality of selves in a plurality of worlds (in which each member of the living world is its own essence-less universe with its own schizophrenic and transient laws of reality and its own particular set of interactions with other members of the world) means that postmodern associationism is inclined to dissolve the 'Humanity/Nature' barrier in a different way from that of the monism of unitarianism.

While unitarian metaphysics dissolves the Humanity/Nature barrier to give rise to one truly great beast called Gaia, the postmodern association view first dissolves Humanity and Nature so that the abstract boundaries that held the individuals (and groups of individuals) together dissipates to allow an unbounded jumble of individuals (and groups of individuals).

If the duality of Humanity and Nature are so dissolved (whereby Humanity and Nature might be considered to have dissolved as well) this would make anthropocentrism a much more complicated beast than the unidimensional monster currently staggering around wreaking havoc upon the world (since there would be no general 'Human' position to be anthropocentric from and no Nature to be anthropocentric against). If we, for the moment, trust the postmodernists and regard Humanity as only a vague and value-laden term used to describe a loose collection of individuals that possess no inherent essence, then the 'anthro' part of anthropocentrism becomes less focused. Indeed if 'Humanity' has been a convenient totalising conceptual tool in the hands of bourgeois, white, male imperialists, then anthropocentrism may be a convenient political tool in the hands of philosophising environmentalists. From such a perspective, anthropocentrism seems to be a unifying concept, uniting Humanity as a common coherently identifiable evil which is homogeneously spread throughout the social world.

Rather than viewing anthropocentrism as a ubiquitous, uniform and singularly encompassing value that inspires attitudes and actions that are detrimental to nature, it is perhaps better presented as a range of value positions which have differing social and philosophical bases and varying types of environmental impact. Anthropocentrism may be just a convenient name (or cover) for, amongst other things:

a) egocentrism (it is not that we do not value anything but humans but that we do not value anything but ourselves);

b) chronocentrism (people are trapped in their own particular historical and temporal framework and they can not see the heritage from which they came nor can they see the future to which they are headed. To many people, their past environments were much as they are now, especially as they are seen through a filter of encultured indifference. If this is so then there is reason to believe that the drastic environmental changes that environmentalists keep going on about are only mild afflictions and the future will not really be any different from the past);

c) humanism (because humans shamelessly elevate themselves "into the absolute masters of creation"[55] they can then go on to rightly act as though they were the ultimate ends of all life);

d) materialism (because material possessions are an indicative measure of one's interests being happily provided for, anything not contributing to that material well-being--for example, most of nature--is valueless);

e) capitalism (alienation from one's own labor, and this applies to both capitalists and labourers, means that one becomes alienated from one's natural heritage, or one's 'inorganic body' in Marxist terms);

f) androcentrism (the way which man dominates nature flows from the way which man dominates women);

g) technologism and economism (due to the reification/deification of technology over and above nature as one of the gods of Modernity, natural objects are now converted into resources which can only be measured in economic terms).

The schizoid multiplicity of anthropocentrism suggests that environmentalists might fight the destruction of the world's biota by addressing a whole plethora of (related and unrelated) value positions rather than engaging in a struggle with one great 'ism' often thought to be lurking at the base of environmentally malevolent attitudes.

Having dissolved anthropocentrism in this fashion, postmodern associationism may still have a specific comment to make in the ethical arena that nevertheless accords closely with some of the non-anthropocentric environmental thinking in contemporary environmental scholarship. As postmodernism respects and celebrates the Other (as a disconnected being from oneself) and as humanity has been dissolved (so that every being on the planet is granted the status of Other, rather than just human beings) and as Nature, too, can be similarly dissolved (so that the members of the ecological world are not represented by a unified concept of collective Otherness) then any postmodern breed of environmentalism might be inclined to celebrate and respect every being with equal fervor regardless what side of the Humanity/Nature barrier it is said to be on. Thus, a non-anthropocentric value system may be allowed to proceed without conferring the totalising and essentalist tendencies of unity upon the ecology of the world.

The Broadening of Social Constructionism

If the ideas of social constructionism present challenges to foundational ideas like 'Humanity' and 'Nature' then we might also find that non-anthropocentric forms of environmentalism present a challenge to social constructionism.

If postmodernists are avid fans of social constructionism, and if they agree to deconstruct the divide between Nature and Humanity, between the natural world and the social world, then we have an interesting thing happening to social constructionism: social constructionism becomes not the preserve of just humans, since the category of Humanity has been deconstructed, it is also the preserve non-humans. All creatures are multiple-selved, identity-perplexed, socially-constructed individuals; whether 'human' or not. This is to say that those regarded as 'non-humans' are also constructors of the stories and metaphors that are formed of them in human culture.[56] Animals and plants are people, too, it could be said--tellers of stories which we have yet to find out how to listen to.

When we admit to the constructionist ability of non-humans, two particular facets of this constructionism might be identified. Firstly, non-humans may help in the construction of stories we have about them, and, secondly, they construct their own stories about the world which are independent of their relationship to our storytelling.

To deal with the first facet: this would suggests that humanity's stories of non-humans, the metaphors that we use to describe them, might, in some way, come from them as well as from other humans. But how is it possible, it may be asked, that non-humans can contribute to human stories about them when the non-humans are not even able to communicate with humans in order to partake in such story construction? Obviously we do not share the same language as non-humans but there are communications going on; albeit fragmented, partial and non-representative conversations (but the same is evident, it might be pointed out by postmodernists like Kroker and Cook[57], in conversations between one human and another).

When I suggest that non-humans partake (in some small way) in the social construction of our stories about them I am not saying that there is some essence, some independent referentiality, that shines through from an animal or plant despite the various human to human social constructions of non-human nature but that our constructions can be influenced by the behaviour of those non-humans. An animal like a bird, for instance, does not reveal itself via its own objective and intrinsic reality into our stories. The bird that we see in our scientific reports, in our novels and poems and myths, is not a mirrored reflection of some independent real bird. It is just that in some of the stories we have about birds, the 'bird-ness' is negotiated into place with the help of the behaviour and activities of birds.

An example of this may be fitting. In a book about bird behavior, Barber describes a sixteen year old male African Grey parrot who has been an experimental bird in the laboratory of German animal behaviourist I.M. Pepperberg for most of his life:

> Alex is not isolated in the laboratory, and he exercises much control over his life. He is out of his cage about eight hours every working day. During this time he interacts with the experimenters, who treat him respectfully, as an important member of the laboratory team, while teaching him words and concepts and testing him in formal, rigorously controlled experiments.[58]

During his sixteen years Alex has shown, according to Barber and Pepperberg, that he is able to communicate an understanding of human conceptual systems, from colour and shape to numbers and sentence structure. Barber is convinced that Alex does not just mimic but actually:

> understands what he is saying, and when he answers a question correctly, he demonstrates confidence in his answer. About 5 percent of the time, student experimenters have mistakenly scolded him for a correct response. Not accepting their reprimands, Alex confidently repeated his correct answer.[59]

What is, of more significance, perhaps, is the following:

> Alex is not passive; to a large extent he actively controls his life. For instance, when a desired nut was placed by an experimenter under a heavy metal cup that Alex could not lift, he told the experimenter 'Go pick up cup'. As soon as the experimenter picked up the cup, Alex walked over and ate the nut. If the experimenters failed to make the experimental tasks sufficiently interesting for him, Alex lets them know he is bored. He communicates his boredom by asking to be moved to a new location or by ignoring the experimenters and preening. During an especially long and difficult experimental session, Alex communicates his frustration and desire to stop answering questions by requesting to return to his cage: 'Wanna go back'. When brought to his cage, he tells the experimenter 'go away', and climbs inside the cage onto the swing, ignoring the experimenter and refusing to interact with any one.[60]

What we could possibly infer in Alex's case is that a non-human can realign the subject of examination and the process of investigation as he or she is being studied. Alex the parrot can contribute to the negotiations about himself, re-sort the problem being examined into new (admittedly anthropogenic) categories, and re-devise the experiment, or curtail its usefulness.

Needless to say this sort of fragmented and partial communication between non-humans and their human observers does not always happen. The voices of very many non-humans are, no doubt, hardly ever heard and so are not able to negotiate in many instances.

To deal with the second facet of the widening of social construction--that non-humans construct their own version of the world--it is here far easier to hypothesise the possible existence of a mediating language between various social constructors, since, for the most part the construction will be within groups of individuals of the same species (and so possess some sort of common evolutionary, physiological, or social heritage).[61] These communications (languages) may be verbal, as in birds, or they may be visual, chemical, tactile etc.

If we are to describe the communications between non-humans as a form of language, this is to implicitly to say that non-humans exhibit awareness, intelligence, foresight, alertness, purposeful behaviour, willfulness, and many other characteristics generally only ascribed to humans. It is also to say that the communications using these languages are involved in conveying maps, telling stories, and exchanging metaphors. Again, perhaps an example is warranted.

Consider the dance language of bees.[62] This familiar tale of communication between the members of a non-human species takes on a new light if we recount a story that Barber narrates about the process of honey bees finding new nest sites:

> After the returned scouts have reported the distance, direction, and desirability of their best nest sites to each other; they all fly out again to inspect sites reported by others. If another bee's find is more satisfactory than her own best find, the scout 'changes her mind' and, when she returns to the swarm, dances for the alternate site. All scouts now literally 'vote with their feet' (by vigour of dancing), and the dancing-voting process continues until they are all come to an agreement.[63]

This example may be a possible instance in which the social negotiation of a contested reality (the best nest site) by a non-human species is undertaken. The way that bees may use their languages to communicate may be entirely unlike the way humans do it. The rules of their exchange, and the processes of the story-telling may be quite unique to the species (or group) that it concerns. A wiggle or a dance is not equivalent to a word, a sentence, or a syntax device in any human language, verbal or non-verbal, but they can nevertheless communicate their own stories to other members of their species whilst these other members offer up their own stories.

A (rather obvious) suggestion that can be made here is that people's reticence to accept that communication/language exists in non-human society is due to the alienness of non-human modes of language and the alienness of the stories they tell. The language of each non-human species is so specific, so accented, so unique, that humans can not recognise the immense variety that might possibly occur in a bird song or a bee wiggle. The scientific stories of humans cannot properly describe all of these different languages and if it could it would need such a vast array of laws, rules and assumptions that it could not be housed within any one theory.

This acceptance of culture within animals does not have to deny the biological and genetic heritage of species but it does downgrade it in importance. This process of downgrading the importance of genetics is a prime concern of those animal behaviorists who describe cultural phenomena in animals.[64] It is from the work of such studies that the sociality and culture within non-humans may be assessed in a new light.[65]

No doubt some will label the views of these people (who see culture in non-humans) as anthropomorphic. However, I would defend these researchers by indicating that they are not declaring that there is any cultural similarities between humans and non-humans only that non-humans *have* cultures: non-determined actions that are the happenstance of historical contingency and experience.

Conclusions: Postmodern Associationism as a Possible Form of Postmodern Ecology

What does any postmodern ecology have to consider? A postmodern ecology would have to give cognizance to many or most of the postmodern characteristics described in Chapter Six.

For instance, a postmodern ecology might try to be anti-scientific. Postmodern associationism might hardly be said to do this. After all, it is a development of Gleasonian community ecology; a tradition that is steadfastly within the tradition of ecological science. However, if we postmodernise Gleasonianism so that it is aware of its own metaphoricity--so that it suspects the realism of science--then postmodern associationism can be made to be critical of its own competence to relay the exact nature of the world.

A postmodern ecology might be pessimistic. Since postmodern associationism is aware of its inability to offer up either an ultimate meaning for the world's existence or a grand theory that describes what is going on in that world, postmodern associationism achieves a certain postmodern affiliation with pessimism.

A postmodern ecology would probably have to acknowledge absence and aporia. The section on *Ungluing the Ecological World* details exactly how absence and aporia, and especially the absence of unity, might be an intrinsic part of both Gleason's ecology and an attempt to postmodernise Gleason's ecology. Where Gleason tried to demonstrate that there is a lack of integration between most members of an ecological community, postmodern associationism goes on to indicate that there is no Invisible Hand, no self-regulation (indeed, no self), and no unifying process holding collections of plants and animals together.

Ecology, if it is to be postmodernised, might also reject the fetishism within Modernism for communication. From this perspective, postmodern associationism can claim to be postmodern if it acknowledges that while communication occurs in the natural world, including between non-humans, such communication is never monosemic; never exact; never entirely trust-worthy. Unlike the communication concepts that ride along within some systems theory (and in the case of Margelef,[66] within systems ecology) a postmodern ecology would acknowledge that communications are polysemous, plural and open to confusion. Postmodern associationism is capable of doing this because it acknowledges the unique and specific experiences of individual members within an ecological community (see the section on *The Postmodern Ecological Individual*). These experiences make any one individual organism liable to act with complete specificity to a common ecological (or social) event.

A postmodern ecology might acknowledge and value pastiche, chaos, heterogeneity and fragmentation. As Burrows explains[67], Gleason envisaged ecological communities as "temporary and fluctuating phenomenon" whose composition was decided by chance confrontations with disturbance, parasitism, geographical specificities and "the availability of species in the surrounding vegetation" (whose composition was determined, also, by chance confrontations with disturbance, parasitism, geographical specificities and by the availability of species in the surrounding vegetation, and so on and so). This results, it might be said, in a chaotic, fragmented, heterogeneous, pastiche of unrelated plant and animal species which possess no inherent ecological integration. Such pastiches are not tailored by any organising factor (such as succession, evolution, or self-organisation), but are chaotically thrown together by innumerable individual circumstances.

If an ecological community is thus disintegrated into its component parts by such a narrative, then it is not possible for an abstract representation, such as the superorganism, Gaia or 'Life' to truly represent the actions of those parts. Each individual has a story which is unique to it, a story which is not workable within a 'general' story of the whole. In this way the postmodern penchant for dismissing representation may be found within postmodern associationism.

Gleasonianism might also be said to reflect the postmodern characteristic of anti-teleology since its reliance on chaos, fragmentation and heterogeneity disavows any effort to indicate that any ecological community is developing in any particular direction. This is demonstrated by the second figure drawn in the section about the metaphors within a postmodernised Gleasonian ecology.

A postmodernising of ecology would confer upon ecology a sense of value in those whose stories and values have traditionally been marginilised; the 'Other'. Any environmental philosophy that recognises the intrinsic value of non-humans might be said to do this. However, postmodern associationism, by taking care not to treat nature as a united Other, can go on to recognise and celebrate Otherness in a much more enduring way than can a unitarian narrative. Postmodern associationism, by involving itself with the widening of both culture (so that non-humans are regarded as being able to possess culture) and of social constructionism (so that non-humans become social constructionists, too) might also unveil a way in which the stories of Others may be subjected to the same intensive quest as is already undertaken in the human sciences.

Postmodernism, as we have seen in Chapter Six, is usually steadfastly involved in the deconstruction of foundational ideas: ideas that are thought to be so basic that they have not been questioned. It is with this spirit of deconstruction in mind that postmodern associationism attempts to both postmodernise ecology and evict from it the foundational ideas that have storied ecology since its inception. In this vein, postmodern associationism seeks to call into question ecological foundations such as the balance of nature, the unity of nature, succession to climax, natural selection, levels of biological organisation, and also the categories of 'Nature' and 'Humanity'. By entertaining the possible dissolution of these foundations, new narratives which reinvigourate the values of difference, dissent, atypicality, meaninglessness and Otherness can be opened up.

Indeed, under postmodern associationism stories are able to be constructed for each and every biological individual that exists in the ecological world. In this way, through atomising the ecological world, the needs, lives, tragedies, interests, values and historical heritage of each non-human may be told in all their variety. Stories which, under unitarianism, are drowned out by the constant re-telling of the one metanarrative that is unity.

Notes for Chapter 8 (*An (Other) Postmodern Ecology*)

1. P. Rosenau, *Postmodernism and the Social Sciences* (Princeton: Princeton University Press, 1992) p.14.
2. Z. Bauman, *Intimations of Postmodernity* (London: Routledge, 1992) p.29.
3. A. Heller and F. Feher. *The Postmodern Political Condition* (Cambridge: Polity Press, 1988) p.1.
4. Rosenau, *Postmodernism and the Social Sciences*, p.81.
5. If no generalisations can be made, laws cannot be found and defined, and theories will always remain gap-ridden and local.
6. Postmodern associationism might be considered an ironic and self-contradictory label for that which it purports to describe in the sense that the members of an 'association' are much of the time not at all 'associated' but merely coincidentally arranged as neighbours in space and time. Gleason, himself, reflects this ambiguity since while using the term association as a key metaphor when describing certain aspects of ecological settings he, in his 1926 paper, alludes to the non-existence of associations as distinct entities:

   "Are we not justified in coming to the conclusion, far removed from the prevailing opinion, that an association is not an organism, scarcely a vegetational unit, but merely a coincidence?" (Quote from H. Gleason, "The Individualistic Concept of the Plant Association", *Bulletin of the Torrey Botanical Club*, 53, 1926, p.16.).

   Having acknowledged this, it becomes acceptable to use the term association as Gleason would have us use it: to refer to a loosely-gathered group of coincidentally-arranged, separate 'unitary' organisms that live in a common (though impossible-to-define) geographical site without contribution to the regulation, maintenance or ordering of the 'whole' contained within that site.
7. R. Tester, *The Life and Times of Postmodernity* (London: Routledge, 1993) p.153.
8. J. Hagen, *An Entangled Bank: The Origins of the Ecosystem Concept* (New Brunswick: Rutgers University Press, 1992) p.123.
9. R. Margalef, *Perspectives in Ecological Theory* (Chicago: Chicago University Press, 1968)
10. T.F.H. Allen T.B. and Starr, T.B. Hierarchy*: Perspectives for Ecological Complexity* (Chicago: Chicago University Press, 1982).
11. E.P. Odum, E.P. *Fundamentals of Ecology* (Philadelphia: Saunders, 1971).
12. P. Davies, *The Cosmic Blueprint* (Harmondsworth: Penguin, 1995) p.149.
13. E. Goldsmith, *The Way: An Ecological Worldview* (NY: Shambala, 1993) p.106.
14. F. Capra, *The Turning Point* (NY: Simon & Schuster, 1982)
15. Recall, from Chapter Six, the anti-hierarchical stances as exemplified by: I. Hassan, *The Postmodern Turn: Essays in Postmodern Theory and Culture* (Ohio: Ohio State University Press, 1987); J. Deleuze and F. Guattari, *A Thousand Plateaus* (Minneapolis: Minnesota University Press, 1987); Bauman, Intimations of Postmodernity. See, also, the discussions on heterogeneity and fragmentation in the same chapter.

16. Yet even this metaphor is unsatisfactory, for it does not capture the idea that the splintered structure of the ecological world emerged from the activities of contextualised individuals but suggests the shards and splinters are mere remnants of a once organised higher whole or level.
17. For instance, this is how Goldsmith looks at reductionism:

    "The Cambridge zoologist and ethologist W.H. Thorpe defines reductionism as 'the attribution of reality exclusively to the smallest constituents of the world and the tendency to interpret higher levels of organisation in terms of lower levels'." (Quote from Goldsmith, *The Way*, p.9).

    This is also the reductionism that Paul Davies thinks he avoids by using levels:

    "Starting with atoms, building up through molecules, cells and organisms to conscious individuals and society, each level contains and enriches the one below, but can never be reduced to it" (Quote from Davies, *The Cosmic Blueprint*, p.146.).

    However, Davies himself is liable to forget this in his own schemes of elaboration, as a sample of his own writing about human sexuality shows:

    "So are we all doomed to act out a pre-programmed genetic agenda that all too often serves to make men and women miserable or can human beings successfully override their genetic legacy? Alas, the statistical evidence suggests that whatever the prevailing social or religious norms, our genes manage to manipulate us with surprising efficiency. We may not be able to alter human sexual behaviour much, but at least it helps to understand how it originated." (Quote from P. Davies, "Basic Urges Take Some Stopping", *The Australian*: HEC Column, The Australian Online--1996).

18. In light of this we witness the convergence of holism and reductionism. If reductionism is that which describes the existence of ecological phenomena according to a set of simple (usually physico-chemical or mathematical) laws, then it stands very close to holism which dictates that ecological phenomena are at the mercy of simplified general laws (which again are usually physico-chemical or mathematical) operating at a higher level.
19. J. Wu and O.L. Loucks "From Balance of Nature to Hierarchical Patch Dynamics: A Paradigm Shift in Ecology", *Quarterly Review of Biology*, 70, (1995) :449.
20. After Kropotkin. Kropotkin's anarchist ideals for utopian societies required the revocation of Darwinian competition and its replacement by a worldview proposing cooperation as the essential driving force of nature. See P.A. Kropotkin, *Mutual Aid: A Factor in Evolution* (London: Wilheinium, 1902). For a modern-day interpretation, see: D. Pepper, *Ecosocialism: From Deep Ecology: Deep Ecology to Social Justice* (London: Routledge, 1993).
21. R.P McIntosh, *The Background to Ecology* (NY: Cambridge University Press) p.266.

22. See for example: R.T. Paine, "Food Webs: Linkage, Interaction, Strength and the Community Infrastructure", *Journal of Ecology*, 2, (1980) :159-170.; M.J. Crawley, *Herbivory: The Dynamics of Animal-Plant Interactions* (Oxford: Blackwell, 1983).
23. See for example: W.H. Drury and I.C. Nisbet, "Succession", *Journal of Arnold Arboretum*, 54, (1973) :331-368.; P.S. White, "Pattern, Process and Natural Disturbance in Vegetation", *Botanical Review*, 45, (1979) :229-299.
24. J.F. Anderson and R. May "Population Biology of Infectious Diseases: Part One", *Nature*, 280, (1979) :361-367; R. May and R.M. Anderson, "Population Biology of Infectious Diseases: Part Two", *Nature*, 280, (1979) :455-461.
25. K. Faegri, and L. van der Pijl, *The Principles of Pollination Ecology* (Oxford: Pergamon Press, 1971); D.H. Boucher *et al*, The Ecology of Mutualism, *Annual Review of Ecology and Systematics*, 13, (1982) pp.315-347.
26. Co-operation gives rise to the community or ecosystem level of organisation by tying all the organisms within a community or ecosystem into a tightly bundled and complimentary assemblage of related parts. Competition ties the individuals to the whole by invoking the concept of evolution through natural selection (which acts as a feedback mechanism that also produces a tight and complimentary assemblage of well-adapted individual organisms).
27. For examples of this, see: See Bauman, *Intimations of Postmodernity*, C. Lemert, *General Social Theory, Irony, Postmodernism*, in S. Seidman and D.G. Wagner, eds, *Postmodernism and Social Theory* (Cambridge: Blackwell, 1992) pp.17-46; P. Waugh, "Modernism, Postmodernism: Gender and Autonomy Theory", in P. Waugh, ed, *Postmodernism: A Reader* (London: Arnold, 1992) pp.189-204.
28. Rosenau, *Postmodernism and the Social Sciences*, p.53.
29. Hassan, *The Postmodern Turn*.
30. On this aspect of postmodernism it might be prudent to remember that "there is often an alternative foundationalism lurking in many postmodern arguments" (Quote from: Waugh, "Modernism, Postmodernism: Gender and Autonomy Theory", p.11.)
31. See W.C. McGrew, "Chimpanzees, Tools and Termites: Cross Cultural Comparisons of Senegal and Rio Mini", *Man*, 14, (1979) :185-214.
32. M. Kawai, "Newly Acquired Pre-Cultural Behaviour of the Natural Troop of Monkeys on Koshua Islet", *Primates*, 66, (1965) :21-36.
33. T. Barber, *The Human Nature of Birds* (Melbourne: Bookman, 1993)
34. See RR. Jackson, "What is that spider thinking?" *New Zealand Science Monthly* (February 1995) :6-8; R.R. Jackson and S.D. Pollard, 1996. Predatory Behavior of Jumping Spiders. *Annual Review of Entomology*, 41, (1996) :287-308
35. L.R. Aronson, "Further Studies on Orientation and Jumping Behaviour in the Gorbiid Fish; *Bathygobus soporater"*, *Annals of the New York Academy of Sciences*, 188, (1971): 378-392.
36. L.L. Cavalli-Sforza, "Cultural Evolution", *American Zoologist*, 26, (1986): 845.
37. V. Shulaev *et al*. "Airborne signaling by methyl salicylate in plant pathogen resistance" *Nature*, 385 (1997) :718-721.
38. S.F. Goldstein, "Flagellar Beat Patterns in Algae", in M. Melkonian, ed, *Algal Cell Motility* (NY: Chapman Hall, 1992) pp.99-145.

39. P. Applewhite, "Learning in Protozoa", *in Biochemistry and Physiology of Protozoa* (London: Academic Press, 1979).
40. Cavalli-Sforza, "Cultural Evolution", p.855.
41. Rosenau, *Postmodernism and the Social Sciences*, p.55.
42. R. Hollinger, *Postmodernism and the Social Sciences: A Thematic Approach* (California: Sage, 1994) p.11.
43. See Goldsmith, *The Way*.
44. If postmodern associationism were a metaphysical announcement of a billiard ball ecology it would state that each of the ecological billiard balls would be of different sizes and weights, would walk about on their own without needing to be hit, would cling unpredictably to other billiard balls, would refuse to sink into the (w)holes, would jump chaotically from table to table and would then probably proceed to invent their own rules of billiards without much recourse to known laws.
45. Birch, *On Purpose*, p.xi.
46. Whittaker and Woodwell, in McIntosh, *The Background to Ecology*, p.239.
47. This general conclusion can be found in the following texts: Goldsmith, *The Way*; Davies, *The Cosmic Blueprint*; F. Capra, *The Web of Life* (NY: HarperCollins, 1996).
48. In the vein of Lakoff, Johnson and Miller. See G. Lakoff and M. Johnson, *Metaphors We Live By* (Chicago: University of Chicago Press, 1980); D. Miller "Metaphor, Thinking and Thought: Part Two" *ETC.: A Review of General Semantics*, 39, (1982) :242-256; D. Miller, "Metaphor and Culture", *Paper Presented to the 1983 Culture Seminar*, University of Melbourne, Australia.
49. Miller, Metaphor and Culture, p.6.
50. *ibid*, p.7.
51. Compare this with the rhetoric of some (self-proclaimed) 'paradigm-shifting' unitarian thinkers who make detailed social and moral prescriptions based upon the universal values that are held to flow from a metaphysical commitment to unity and systems thinking (see B. Devall and G. Sessions, *Deep Ecology: Living As if the Earth Really Mattered*, Layton: Gibbs M. Smith 1985; C. Spretnak, *The Spiritual Dimension of Green Politics*, Santa Fe: Bear, 1986; Birch, *On Purpose*, Goldsmith, *The Way*; D. Zohar and I. Marshall, *The Quantum Society: Mind, Physics and a New Social Vision*, London: Flamingo, 1994.
52. R. Hollinger, *Postmodernism and the Social Sciences: A Thematic Approach* (CA: Sage, 1994).
53. For example, see T.F.H. Allen and T.B. Starr, *Hierarchy: Perspectives for Ecological Complexity* (Chicago: Chicago University Press, 1982), Bazzaz, F.A. and T.W. Sipe "Physiological Ecology, Disturbance, and Ecosystem Recovery", in E.D. Schulze & H. Zwolfer, eds, *Potentials and Limitations of Ecosystem Analysis* (Berlin: Springer-Verlag, 1987); pp.203-227; R.G. Wiegert "Holism and Reductionism on Ecology: Hypothesis, Scale and System Models", *Oikos*, 53 (1988):267-669; H.J. Carney, "On Competition and the Integration of Population, Community and Ecosystem Studies", *Functional Ecology*, 3, (1989):637-641, L. Westra, *An Environmental Proposal for Ethics* (Baltimore: Rowman and Littlefield, 1994); Wu and Loucks "From Balance of Nature to Hierarchical Patch Dynamics"

54. Bauman, *Intimations of Postmodernity*, p.xxiii.
55. Levi-Strauss quoted in D. Pace, *Claude Levi-Strauss: Structuralism and Sociological Theory* (Boston: Routledge and Kegan Paul, 1983) :247.
56. For ease of expression, allow me, here, to continue to use the terms 'humans' and 'non-humans' in the following pages, even though I have just dissolved the categories they belong to.
57. A. Kroker and D. Cook, *The Postmodern Scene: Excremental Culture and Hyper-Aesthetics* (London: MacMillan, 1991)
58. Barber, *The Human Nature of Birds*. See, also: I.M. Pepperberg, "Cognition in the African Grey Parrot: Preliminary Evidence for Auditory/Vocal Comprehension of the Class Concept", *Animal Learning and Behaviour*, 11, (1983):179-185; I.M. Pepperberg "A Communicative Approach to Animal Cognition: A Study of the Conceptual Abilities of the Grey Parrot", in C.A. Ristau, ed, *Cognitive Ethology: The Minds of Other Animals* (NJ: Lawrence Erlbaum, pp.153-186).
59. Barber, *The Human Nature of Birds*, p.7. Barber assures us that the African Grey parrot is, by all accounts, living a happy life of both contentment and adventure within this laboratory.
60. *ibid.*
61. The acceptance of pervasive intra-species communication (language) between non-humans might not be forth-coming from every reader but it would be a brave and fool-hardy person to claim that there has never been, in the whole history of the world, at least two members of a non-human species that have managed to communicate with each other in some instance
62. As described for instance in M. Lindauer, *Communication Among Social Bees* (Cambridge: Harvard University Press, 1961); K. Von Frisch, *The Dance Language and Orientation of Bees* (Cambridge: Harvard University Press, , 1967) or C.D. Michener, *The Social Behavior of Bees* (Cambridge: Harvard University Press, 1974).
63. Barber, *The Human Nature of Birds*, p.144.
64. B.B. Beck, *Animal Tool Behavior: The Use and Manufacture of Tools by Animals* (NY: Garland STPM Press, 1980); J.T. Bonner, *The Evolution of Culture in Animals* (Princeton: Princeton University Press, 1980); L.L. Cavalli-Sforza, "Cultural Evolution". *American Zoologist*, 26, (1986) 845-855; I.M. Pepperberg, "A Communicative Approach to Animal Cognition: A Study of the Conceptual Abilities of the Grey Parrot"; K. Pryor and K.S: Norris, eds, *Dolphin Societies and Puzzles* (CA: University of California Press, 1991); C.E. Ristau, ed, *Cognitive Ethology: The Minds of Other Animals* (NJ: Lawrence Erlbaum, 1990).
65. Incidentally such a program of incorporating postmodern cultural studies into biology rallies behind the call by Jagtenberg and McKie to have cultural studies evaluate environmental issues. See T. Jagtenberg, and S. McKie, *Eco-Impacts and the Greening of Postmodernity* (CA: Sage, 1996).
66. Margelef, *Perspectives in Ecological Theory*
67. Burrows, *Processes of Vegetation Change*, p.432.

SUMMARY

If asked to identify those cultural concepts that have given rise to deleterious environmental worldviews, and therefore to deleterious environmental practices, the concepts of atomism, mechanicism, reductionism and dualism--in all their various guises--are said to be the most dangerous. The antidote to this quartet of death and decay, it is said, is an enculturation within science (and within society) of three metaphysical saviours: holism, organicism and, most especially: unity. Unity will unite all the processes and objects in the universe under a worldview that might advise us how to understand the various parts of the universe. Unity will also unite us in our attempts to revere nature and the things in it, most notably, by classifying nature, and the things in it, as alive and therefore in some way valuable. Unity will also offer a perspective of long-termism, overallness and universality from which social processes might be derived and enforced so that we do not destroy nature and ourselves.

The conclusions of this book, however, suggest something different. Not only is unity a socially dubious concept that plays into the hands of conservatism, managerialism, and technocentric imperialism but when unity narrates about the reality of the terrestrial ecological communities of the world, which is one of the prized sites of interest for environmentalists, it also becomes a totalising, foundational idea that inappropriately distributes value within those communities.

When looking at society and nature it is often the case that social scientists lambaste those who take any thing from biology and apply it to society. This work does not adhere to this view although it is sympathetic with it. It adheres to a view that says that we cannot get away from speaking metaphorically in everything we say about nature and society and that no referent (including the big ones like Humanity, Nature, Unity, Truth etc) can be referred to without some form of metaphorical utterance. In this case each metaphor must not be castigated as inappropriate just because it comes from nature to society or from society to nature since all our speeches and writings about nature and society are metaphors that probably contain within them references to things in both nature and society. Each metaphor thus has to be judged on its own merits with reference to its peculiar political aspirations and for the things that it does in that context.

If metaphors are omni-present in our modern-day descriptions of the world, and probably in all our daily lives, we might like to choose to investigate the original referent from which a metaphor was drawn so that we can better see if it is politically appropriate or not. However, it seems that in the case of the metaphors examined in this work, and probably in the case of all metaphors, no point of origin and no original referent can be determined. Because no point of origin can be determined for any particular metaphor then the original site can hardly claim to be absolutely original. The original referent no more adequately reflects reality any more exactly than the metaphors which might have descended from it.

Theorising about the natural and social worlds means the rhetorical re-activation and re-deployment of metaphors (in the vein of Mirowski's 'reprojective spiral narratives'), none of which can maintain purchase on an original referent; they merely slip and jostle passed one another in their bids to be believed and/or utilised. Neither nature, nor society, are the base from which things were originally extracted. Each feeds into the other and back again.

This inability to escape metaphors in our descriptions of the world (and if Lakoff, Johnson and Miller are to be trusted, in our daily activities, speeches and acts) seems to leave us at odds as to how to locate ourselves and our politics since both ourselves and our politics are marked by an 'irreferentiality'; an absence of the real from which to concretely generalise, philosophise and politicise. This does not however preclude us from making a sustained critique of, and raising an alternative to, any one metaphor since we can acknowledge that within their own schemes of elaboration, transmission and value, a metaphor makes connections with other metaphors that may or may not be self-consistent with the declared aims of the users. In the context of this work this is to state that unitarianism is a metaphor whose links with other metaphors undermine the very politics and values of its elaborators. Unitarianism overwhelms its advocates' particular political positionings by its association with other metaphors. This, accompanied by its claim to be a non-metaphorical natural truth (and also the only ecopolitically-valid philosophy of nature) suggests that unitarianism is not only vulnerable to colonisation by anti-environmental thought but that it seeks to overwhelm all thought about nature in contemporary times.

As we have seen in Chapter One, many philosophers of nature (who double as scientists or environmentalists) are nominally aware of the metaphorical nature of large-scale human pronouncements about the world. For instance, people like Goldsmith, Birch, Davies, Ferre, Merchant and Capra are of the opinion that mechanicism is a vast metaphorical story with grand deleterious repercussions for the environment. Notwithstanding the critique in Chapter Seven (which declares the mechanicism versus organicism division to be artificial), these writers are concerned that the metaphorical nature of the mechanistic worldview has become such a dominant philosophy of nature that scientists and science managers are very often unable to escape the consequences of using these metaphors; positing a dead Earth, reinforcing anthropocentrism, establishing fragmented policies and research programs, and devaluing the environment and its members. For the writers listed above, a new view of the world is emerging that renders mechanicism merely an outdated and fallacious metaphor. This new view is that of holistic unity, organicism and self-organising complexity.

Environmental thinkers and natural philosophers are not the only ones involved in the construction of 'new' scientific worldviews, however, and in Chapters Two, Three, Four and Five it is suggested that environmentalists have, in a major way, had their view of the cosmos constructed for them by ideologues of a technocentric, conservative and capitalist bent. The metaphors used by environmentalists and environmental sympathisers like Goldsmith, Capra, Zohar, Birch, Devall, Sessions, Merchant are the same metaphors used by capitalist and conservative apologists such as Hayek, Merry, Ayres, More, Maley, Stacey and Rothschild.

In Edward Goldsmith's case we find that he promotes the unity of nature idea not just for environmental reasons but also for promoting a kind of conservatism that naturalises stability and changelessness in the social world and which reifies hierarchy and consensus in favour of heterarchy and dissent. Although the labels that Goldsmith bandies about (labels like 'the organic worldview' and 'the holistic paradigm') make his philosophies sound very eco-friendly, it is concluded here that such organicism and holism does not do the things for the environment that Goldsmith believes. Goldsmith (and the other holist philosophers of nature critiqued in this book) suggest that an organic holistic worldview vitalises what Cartesian science has killed; it supposedly gives life to non-human members of the universe and it gives life to the Earth. However, in the case of those scientists that are taken as paradigmatic exemplars of this new organic/ecological worldview such conclusions are not valid. The life that Eugene Odum' s systems theory gives to ecosystem constituents and the life that Lovelock's Gaia theory gives to the Earth is a shallow physico-chemical kind of life. So shallow is this version of life, since it is measured in the materialistic terms of energy and matter (and how Descartes would have loved that), that the members of the ecosystem and the biosphere display no other sign of life apart from their interactions with energy and matter. Gaianism doesn't give life to world, it sucks the life out of living things. Any remaining value is then distributed around the natural world in strict accordance with the functional attributes and niche roles that underlie the transference of energy and matter.

We also find that in the end what Goldsmith, Davies and Capra are advocating as holistic science might also be more appropriately positioned within the realms of its perceived opposite: reductionist science. Where Goldsmith, Davies and Capra would advocate Clements', Odum's and Lovelock's superorganismic theories as holistic, we might just as well say superorganicism is reductionistic since Clements and Odum were attempting to physiologize ecology, since they tried to reduce ecological situations to physiological systems. For Odum (and his systems ecology colleagues) this physiologising of ecology meant reducing the members of an ecosystem into mere shells; robotic components that were in existence only in order to transfer what were thought to be the real essences of the natural world: energy and matter. In this light it is not necessary to envisage holism and reductionism as opposites, nor are they two sides of the same coin. They exist--at least in ecology and environmental science--within each other. Holism attempts to characterise the whole gamut of activities within a field of interest by observing common emergent patterns which are then seen to be operable according to some simple generalisation. Reductionism similarly attempts to offer a simple generalisation of an acknowledged group of collective activities.

In both holism and reductionism we thus find very similar theoretical consequences. Where reductionism in biology and ecology reduces the actions of populations, communities and ecosystems first to population and organismal genetics and thence to molecular biology and organic chemistry and thence to the mathematical principles of a very few universally-applicable scientific 'laws', so holism first reduces the activities of populations, communities, ecosystems to a general set of emergent properties which are held to be workable entirely within the narrow and simplified mathematical 'laws' of self-organisation and self-regulation

Gaia theory, perhaps the grandest holist/reductionist theory of all, possesses many of the problems that false holisms exhibit. It is not only implicitly predisposed towards environmental endangerment by positing a value framework that denies the value of individuals and individual species, it is explicitly anti-environmental in the social sphere; positing economic growth, high technology, and interplanetary colonisation as part of its scientific/ideological framework. These explicit implications are not attached to Gaia like unimportant redundant organs. Gaia was made up and developed and pursued by James Lovelock precisely because it sanctioned and sanctified this framework. Lovelock is also of the belief that other people should adopt only his particular reading of Gaia too. He endlessly berates 'wrong-headed environmentalists' while promoting functionalism, economic growth, big science, nuclear technology and space exploration in his writings on Gaia. Here we might note that if we are to take seriously the deconstruction of anthropocentrism that is attempted in Chapter Eight, we might want to confirm that Gaia isn't really anthropocentric, but Lovelock-centric. All those whose ideas coincide roughly with Lovelock's thus suffer from the same kind of centrism; believing that their Gaian Goddess is advocating--through her sacred unchanging laws--a world of 50 million white temperately-situated middle class people whose natural right it is to possess high technology and nuclear power stations and who quite naturally express a suitable desire for economic growth and interplanetary conquest.

Just as Gaianism (and systems ecology and Clementsian plant ecology) give rise to very obvious ecological fascism--where the value of individuals are sacrificed for the good of an abstract whole--so it has been charged by other writers, notably Janet Biehl and M.E. Zimmerman that the 'unity of nature' idea possesses potential socially fascist ideas. Whether or not this is so has not been resolved in this book. However a more important point is raised. Unitarianism might give rise to social fascism of the Nazism type but the embodiment of the 'unity of nature' idea in the form of the 'system' concept is the way in which this fascism manifests itself in contemporary society via the much more politically agreeable thought of liberal capitalism. If the ultimate opprobrium for postmodernism is totalisation then the most common manifestation of that totalisation is the system concept. Radical politics must fight not just against 'The System' (the status quo system of whatever government) but also against the concept of 'system'. By the time we have called the amorphous thing that political activists are fighting against a 'system', the system has already one, since we have already accepted the reality, order and necessity of those things which functionalise us, detract from our individuality and deny us indeterminism.

If the systems theory within systems ecology and Gaianism seems to be a legitimisation of certain forces in modern-day industrialism, we must also note that its child; 'complex systems theory', is also the legitimiser of such conservative forces. Not only that, though, complex systems theory seems to be a legitimiser of one particular social form within modern industrialism: Free-Market capitalism. This legitimisation is not half-hearted. Firstly it involves the naturalisation of the Free Market through the 'order from chaos' ideas of the New Sciences and then it involves the deification of the Free Market when it is claimed by various New Scientists (most notably Capra and Davies) that anything self-organising is a manifestation of God. It seems now that it is not enough to present the Market as an all-pervasive natural process but as an actual God whom we must revere, learn from and never question. If the New Scientists have their way and give rise to a society literate in the importance of this self-organisation, and if Capra and Davies have their way so that humans deify self-organisation, then the future of natural philosophising discourse may tend towards tyranny. Anybody caught interfering with the self-organisation of the Free Market will not only be thought of as going against nature but against God.

According to the New Sciences, God--through the magnificent and divine transcendental process that is self-organisation--has, like many other divine beings, our 'overall' best interests at heart. He supplies our needs if we revere Him and act according to His will. (In economics and society, by the way, this means the preservation of those things which enable the Smithian Invisible Hand to operate). If we do not, we will wreak havoc upon ourselves and our environment.

The Godliness of self-organisation also manifests itself in other ways. The Goddess that is Gaia, for instance, has her God-hood bestowed upon her in such a way by Gaians that we are compelled to believe that she is the ultimate Earthly example of self-organisation. Bowing to Gaia, the Earth Goddess, is not only to submit to the dubious environmental values of Gaianism, however, it is also (as suggested in Chapter Three) to celebrate the Gaian potential to spread human destruction throughout the Solar System.

If science and technology give rise to such metaphysical legitimisation of environmental destruction and social malevolence then what are socially and environmentally-concerned holistic philosophers of nature to do? Well, one of the things they have done is to flirt with the broad and fragmented critique of science as put forward by the intellectual movement known as postmodernism. This strategy is flawed, however, as the only thing of consequence that emerges from this flirtation with postmodernism is a name for their philosophy of nature: 'Postmodern Science'. Postmodern Scientists have successfully latched on to the term 'postmodern' but that is all they have latched on to. Most of the ideas and critique of scholarly postmodernism completely eludes them. And, as is outlined in Chapter Six, the actual metaphysics behind Postmodern Science is patently similar to that of the systems theorists and New Scientists. This metaphysics itself, as we have seen, cogently legitimises what the Postmodern Scientists often make out that they are trying to delegitimise; authoritarianism, Liberalist capitalism, and imperialism. Not only that but it seems as though Postmodern Scientists are involved in an imperialist conquest of their own; to colonise all disciplines of thought and practice and bring them into the fold of Postmodern Science.

The conclusion that comes from looking at Postmodern Science closely is that Postmodern Science is not postmodern. Postmodern Science contributes to the Modernist ideals of unity; hierarchy, overlording holism; and intellectual totalitarianism. Postmodern Science, like the New Sciences, claims victories over reductionism and mechanicism but these are shallow victories since the mechanist/organic divide, and the reductionist/holist divide are artificial divisions with only superficial relevance for environmental evaluation. Buying into these divisions as it does, Postmodern Science only succeeds in extending the tyrannies contained within them: abstract wholes are valued over multiple-selved individuals; systems are reified and deified over the parts they profess to organise; intrinsic value is extinguished in favour of functional value; order is valued over chaos and 'lives' are turned into a single lifeless 'Life'. The detour of Postmodern Science is thus a blind alley.

If Postmodern Science has so obviously failed to be postmodern then where do we go next?. Well, as Postmodern Science attempts to be a predominately 'constructive' narrative of the universe, might we not force our way into the diametrically opposed direction? Instead of pursuing 'constructive' postmodern ideals, shouldn't we try 'deconstructive' postmodern ideals so that we might effect a creative disintegration of the living world? Rather than directly and actively building a metanarrative on the basis of previous metaphysics, might we not just undertake a radical deconstruction of those metaphysics to see what we arrive at? Within the remnants and crumpled ruins of Modernist scientific conceptions of ecology, can we find viable (albeit necessarily fractured) postmodern narratives? Perhaps. And maybe one such narrative is 'postmodern associationism'.

Postmodern associationism is not an enlightened ecological vision that examines all of nature to expose its absolute truth but merely the broken, heterogeneous, fragmented, localised remains that arrive from investigating the blind spots of Modernist metaphysics. Such a narrative is reactionary (reacting against holism in all its forms), such a narrative is socially-constructed (through science, ecopolitics and postmodern theory), such a narrative (despite being applicable to only to terrestrial forest communities) might even be tethered to yet-to-be investigated and not-so-innocent foundations. But it is still a narrative that releases the stories of the members of these terrestrial ecological communities from an imprisoning unity, and, unlike the literal interpretations of the unitarians, postmodern associationism is a metaphor that admits its own metaphoricity.

When we arrive at the associationist version of a postmodern ecology we find that the claims recounted in Chapter One about atomism being a heinous conception of nature may not be valid. Atomism is not such a deleterious natural philosophical principle as holists think, and it may even be a substantial contributor to an environmental ethics respecting Otherness.

We also find that when deconstructing Nature and Humanity we have come more than just full circle with respect to the social constructionist approach alluded to in the Introduction. Where environmentalists and philosophers of nature urge us to study nature on its own terms so as not to destroy it, and where social constructionists (and especially the postmodernists) believe that such studies are socially constructed, this work suggests that social constructionism must anticipate the widening of its own boundaries in the face of the deconstruction of 'Nature' and 'Humanity' so that all living individuals come to be seen as social, and therefore as social constructors. A project to re-story the living world in this light might serve to creatively magnify the value of all the Earth's members by providing a new array of diverse metaphors and ideas.

GLOSSARY

This glossary serves two purposes. Firstly, it defines and introduces specialized terms which may be unknown to readers hailing from diverse disciplines. Secondly, it outlines the particular way that a general term is utilised in the main text.

anthropocentrism: the cultural or philosophical predilection towards regarding humans and humanity as the greatest or only source of value in the natural world.

anthropogenic: that which is made by humans.

aporia: the lack of channels between one thing and another.

association: the term used by ecologist Henry Gleason to describe a patch of vegetation.

atomism: the view that the world is composed of distinct bits and pieces, usually quite small ones.

autopoiesis: a word coined by the Chilean neurobiologists Maturana and Varela and which is synonymous with the term self-organisation. Translated directly from Latin, autopoesis means self-making. Autopoietic networks are networks that make themselves.

autotroph: an organism (generally a green one) which makes it's own food (generally through photosynthesis).

biocentrism: the ethical view that all members of the biotic world, from ameobae to humans, should be valued at some sort of parity

carbon-fixing: the biological process whereby the carbon which exists in the air as carbon-dioxide is fixed into the structural make-up of an organism. Green plants do this particularly well when they take in air through their leaves and turn it into all manner of amazing substances, a well known example being wood.

Clementsianism: the ecological theories of Frederick Clements, including succession theory, the superorganism idea and the theory of climax communities.

chaos: traditionally chaos is defined as that which is random, disorderly and unpatterned. In contrast to such a definition, Chaos theory suggests that chaos is not random and disorderly but is, in fact, a very complex type of order.

Chaos theory: a modern-day theory of chaos derived from systems theory which states that within chaos their is an implicit, though very complex and often unpredictable, order.

Chaotician: a practitioner of chaos theory.

chaotic attractors: when the phenomena studied by Chaoticians is represented graphically on grid paper, a mildly interesting recurring pattern sometimes emerges. The mathematical point about which these patterns revolve is given the name chaotic attractor. However, it is not silly to regard chaotic attractors as phenomena restricted to such graphic representation since that is where they are studied and analysed into existence.

chloroplasts: the organised bits within a plant cell that do the photosynthesizing. They are usually green due to the presence of the famous molecule chlorophyll.

climax: The supposed teleological endpoint of an ecological community that has undergone succession.

community ecology: a branch of ecology that deals with the interactions between various animals of plants in a given area. Community ecology is generally held to exist as a separate field from both ecosystem ecology and population ecology.

complexity theory: a component theory of complex systems theory which deals specifically with the arrangement of complex phenomena in the universe.

complex systems theory: this is systems theory made more complex and dynamic in order to deal with more complex and dynamic phenomena. Amongst the theories that make up complex systems theory are Chaos theory, complexity theory and self-organisation theory. A number of complex systems theorists refer to their field as the New Sciences.

constructionism: (social constructionism) a philosophical and sociological approach to studying humans (and their intellectual and technical products) which emphasizes the importance of focusing on the social processes we all use to create and maintain our beliefs about the world. Under such an approach, scientific facts and theories are held to be beliefs since they can be shown to be constructed and shaped by people more in response to what other people are saying and doing than in response to what some independent reality, like nature, is saying and doing.

control theory: the main theoretical component of the science of cybernetics. It is more or less synonymous with feedback theory and deals with feedback mechanisms in a myriad of phenomena, from electronics and machines to the physiology of cells, organisms and ecosystems.

cybernetics: The science of feedback theory and control theory. Cybernetics composes a major part of systems science. Those who practice in the science of cybernetics call themselves cyberneticists or cyberneticians.

deconstruction: a piece-meal unveiling of the various assumptions in a theory by reworking the historical and sociological processes that made that theory. Deconstruction is often what adherents to constructionism say they are doing.

dissipative structures: structures held to be resulting from the running down of a system but which, paradoxically, seem to involve increasing order and structure.

Deep Ecology: a non-anthropocentric brand of environmentalism that relies on the personal experiences of human individuals with nature to effect environmental attitude changes within society. Deep Ecology‘s deepness has been variously attributed to its deeper respect for nature, its deeper perspective of (biotic) egalitarianism, its deeper analysis of environmental problems and its deeper affiliation with things spiritual.

determinism: where there is a certain inevitability of certain actions and effects. It is the scientific or philosophical equivalent of divine preordainment where things are set in motion to achieve certain outcomes.

downward causation: a term used to describe the effects that higher (larger) levels of organisation have on lower (smaller) levels of organisation. The use of this term implies that there is, in nature, a process whereby lower levels of organisation, say cells and bodies, contribute to the making up of higher levels of organisation, say ecosystems, which then effect a downward influence back on the smaller level.

ecological community: a unit of study in ecology which includes all the living members of a designated area of study; be that area a large stand of trees, a small lake or a community of microbes living in the cavity of a human body.

ecology: the science that studies groups of organisms and the relationships they may make between each other.

ecosystem: a term used to describe a natural area whose living and non-living parts are united through physical interaction; primarily the exchange of matter and energy.

ecosystem ecology: the branch of ecology that studies ecosystems.

emergent properties: those properties that emerge from the interaction of various parts in a collection of objects. Emergent properties can only emerge from such interactions since their various parts do not or can not exhibit such properties by themselves.

evolution: a term used to describe some sort of progressive development.

extant: species which are still living (i.e.: the opposite of extinct species)

feedback: the term used to describe a situation in which a process produces a certain outcome which then feeds back to influence the original process. The principle of feedback is a core component of cybernetics and control theory.

foundation: in scholarly matters, a foundation is a large-scale idea that underpins small-scale ideas. Many foundations are of such a fundamental and obvious nature that they usually go unquestioned. Foundational ideas claim to be capable of situating, characterizing and evaluating all other ideas since they, themselves, are essential truisms agreed upon by almost everyone.

foundationalism: term used to describe the situation whereby scholars structure their ideas and arguments upon fundamental precepts and commonly-accepted metaphysical frameworks. Postmodernists tend to try and eject foundationalism when they partake in the scholarly process of deconstructing Modernist ideas. However, whether they ever stand a chance of being successful or not, is debatable.

fractal geometry: a brand of geometry dealing with fractals; self-similar repetitions of particular patterns.

functionalism: a point of view which describes and studies a social or ecological setting according to the functions and roles of various components within the setting.

Gaia theory: a theory that states that the Earth functions somewhat like a living organism; made up of various parts whose interactions create a self-regulating stable whole.

geophysiology: name given to the science associated with the Gaia theory by its founder, James Lovelock.

Gleasonianism: name used to refer to the botanical theories of Henry Gleason, whereby ecological settings exist as random, anarchic, ill-defined associations.

herbivory: the predation of plants.

hermeneutics: the philosophical study of meaning.

holism: in both science and social science, holism refers to the desire to understand by studying nature or society as whole systems, rather than the desire to understand nature or society by studying just their parts.

homeostasis: the tendency of a system to resist change and disturbance and to maintain itself at some sort of constancy or balance.

levels of organisation: this term refers to the traditional way of viewing the arrangement of entities and systems in the natural world. It is the contention of this book that levels of organisation only exist in the abstract.

limnology: the scientific study of life in lakes and rivers.

materialism: the philosophical approach that studies nature purely by looking at its material elements. It is roughly synonymous with substantialism

mechanicism: the view that all things are basically of a mechanical nature; existing like machines and explainable as such.

metanarrative: if a narrative is a story that seeks to explain something then a metanarrative is a large-scale story that seeks to explain everything. Metanarratives thus have the habit of consuming and subsuming smaller stories while never having their own narrative structure questioned to any great extent.

metaphor: the description of one thing in terms of something else. This would mean that analogies, similes, synecdoche, and even opposites, are brands of metaphor, since all of these things describe one thing in terms of some thing else.

mitochondria: the organs in cells which do the respiring, releasing energy from some food source by attaching oxygen to it.

Modernism: a catch-all term used by postmodernists and others to describe the main intellectual ideas promulgated within the Western World since about 1850. These ideas include Freedom, Truth, Reason and Mankind. A salient feature of Modernism is said to be its adoption of a mechanical, atomistic worldview which came from Nineteenth century techno-scientific re-workings of Descartes' and Newton's 'rational' approach to nature.

New Sciences: a catch-all phrase that refers to a group of scientific theories which includes Chaos theory, complexity theory, self-organisation theory and Gaia theory.

niche: The role a species plays within its ecological setting.

non-anthropocentrism: a value system which in some way holds that humans are not the only things of value in the universe. There are many breeds of non-anthropocentrism, one of the most commonly acknowledged being biocentrism.

Other: a term used to indicate an unchosen thing or collection of things which, for one reason or another, can not be classed as belonging to a chosen category of things. In the social sciences it has become common practice to examine how various groups such as women, blacks and homosexuals, have been 'Otherised' throughout history; castigated as unimportant exceptions to mainstream humanity.

order: term used to describe objects, processes and events which are patterned, organised, well-arranged and able to be utilized as such. This book suggests that there may be no such things in the natural world (especially when it comes to the ecology of the Earth's biotic communities).

organicism: a paradigm within science, philosophy, and environmentalism which states that the universe, and all the things in it are somewhat like living organisms.

paradigm: a paradigm is a generalized way of seeing the world (or a particular section of that world), a way which in turn effects those things you do in that world (or that particular section). A scientific paradigm is the way science sees the world, or more particularly, the way a particular community of scientists see that section of the world that is of interest to them. A social paradigm is a way the general members of society see the world. In this sense 'paradigm' is roughly synonymous with the term 'Worldview'.

photon: a particle of light.

physicalism: A philosophical predilection towards describing all the world's various events and happenings in physical terms. 'Things are things!', physicalists might say.

physiology: The study of function and structure within the bodies of animals, plants and microbes. Some people, for instance, Frederick Clements, Eugene Odum and James Lovelock, hold that structures larger than individual bodies, such as ecosystems and the biosphere can possess a physiology and so can be studied using the principles of physiological science

population ecology: branch of ecology that specializes in the study of populations. In ecology a population is a group of individuals of the same species.

postmodernism: an intellectual movement whose origins are many and whose main positions are multifarious. Generally postmodernism criticizes those foundational assumptions held within Modernism, such as Truth, Freedom, Mankind and Progress. The economic, social and political manifestation of postmodernism is usually termed postmodernity. Postmodernism might easily be confused with Postmodern Science but they are not related in any strong way.

Postmodern Science: Name given to a specific paradigm or worldview that posits that the universe consists of dynamic, interlocking self-organising systems. These systems, both separately, and together, are held to exhibit a large degree of very complex order. Such a worldview represents, it is said, a post-Cartesian and Post-Newtonian paradigm. Postmodern Science is roughly synonymous with the New Sciences. It would be natural to think that Postmodern Science is directly related to postmodernism but it is suggested in this book that they are not strongly connected.

Progress: advancement in a certain direction with an implicit judgment that things are getting better or more complex. Within the Age of Modernity, economic, technological, scientific, social and moral progress were often thought to be constantly present. Within postmodernity, such progress is seriously doubted.

realism: the philosophical view that there is a real world out there and that it contains things that we can know and understand. In this book realism is contrasted to constructionism

referent: an actual thing out there in the real world. Some people argue that referents are more or less fully knowable and some people argue that they might exist but we can not know them. Adherents to postmodernism often argue that there are no referents at all and when we think we are talking about actually-existing real things out there, we are merely making up some sort of story.

re-presentation: this term alludes to the impossibility of representation. When someone represents someone else or when we say something represents something else they do so only via various social, cultural or metaphorical filters which act to actually re-present (adjust, adapt) the original person or thing. Representation, thus, necessarily involves re-presentation.

reductionism: the process whereby a myriad of phenomena or ideas are reduced to a single (or a very few) theoretical principles or laws.

self-organisation: this is the making of a self by the self. In the New Sciences such self-making is a ubiquitous phenomenon in the universe and applies at the scale of cells, organisms, ecosystems, societies, planets and galaxies. The term self-organisation is synonymous with the term autopoiesis.

self-organisation theory: a body of theoretical work that deals with the phenomenon of self-organisation throughout the many realms in which it is found. Self-organisation theory is one of the core theories of the New Sciences

self-regulation: the regulation of the self by a self. When something is self-regulating it indicates that a self is maintaining its own integrity and status. In this book the term 'self' is used to identify a coherently united entity which exists in space and time.

substantialism: a general philosophical idea that alludes to the omni-presence of substances in the world. Extreme substantialism would posit that the world is made up of nothing but substances, variously arranged.

structural functionalism: a general theoretical approach within sociology and political studies which identifies different structures within society and then postulates how these structures function to create a the social world we know.

succession: an ecological theory which states that there is a progressive change in animal and plant life on any newly created geographical site which proceeds from initial colonization to a final stable endpoint called the climax.

symbiosis: the ecological situation whereby members of different species are involved in close relationships with each other to enable them to survive (or to increase their chances of survival).

synecdoche: literary process by which the whole of a thing is substituted for a part, or, conversely, a part is substituted for a whole.

system: a complex of connected things or processes that make up a unity. Most systems are held to be organised, controlled and maintained by feedback.

systems ecology: a branch of systems theory which studies the systems within ecological settings.

systems theory: The science that studies systems. A central principle of systems theory is feedback.

technocentrism: a brand of environmentalism which holds that the environmental crisis can easily be solved within the normal technological and economic programs pursued by most governments.

teleology: this is the practice of explaining events and process in nature or society in terms of the final end-state that they are said to be working towards. Design, determinism, evolution and pre-destiny are often referred to in teleological studies.

totalisation: to totalise is to presume you have worked everything out, to assume that your one cherished idea/theory can explain away all small disagreements. Examples of those that have suffered from, or enjoyed, totalisation in the past are Adam Smith, Karl Marx, Thomas Locke, Immanual Kant, Isaac Newton and Charles Darwin. In all of Modernist thought there seems to be a strong ongoing impulse to Totalise. Postmodernists love picking on totalisers because it gives them an excuse to enter into deconstruction.

unity: a united entity composed of non-separable parts which act in a unified, integrated and interdependent manner (whether conscious of it or not) toward a common agenda: the maintenance of the unit-entity as a whole. Usually, the parts in a unity are held to be interconnected to all other parts.

wave-particle thesis: the idea that light is both a wave and a particle depending on how you measure it.

BIBLIOGRAPHY

Abercrombie, N. *et al* (1984) *The Penguin Dictionary of Sociology*, Penguin, London.

Abraham, R. (1994) *Chaos, Gaia, Eros: A Chaos Pioneer Uncovers the Three Great Streams of History*, Harper, San Francisco.

Abraham, R. and Shaw, C.D. (1988) *Dynamics: The Geometry of Behaviour*, Vol 4, Aerial Press, Santa Cruz, California.

Abram, D. (1985) The Perceptual Implications of Gaia, *The Ecologist*, 15, 3, 96-103.

Abram, D. (1992) The Mechanical and the Organic: On the Impact of Metaphor in Science, *Wild Earth*, 2, 2, 70-75

Abram, D. (1997) Returning to Our Wild Animal Senses, *Wild Earth*, 7, 1, 7-10.

Adorno, T. and Horkheimer, M. (1947) *The Dialectic of Enlightenment*, Cumming, NY.

Allaby, M. (1989) *Guide to Gaia*, Optima Books.

Allen, T.F.H. and Starr, T.B. (1982) *Hierarchy: Perspectives for Ecological Complexity*, Chicago University Press, Chicago.

Allen, T.F.H. and T.W. Hoekstra. (1992) *Toward a Unified Ecology*, Columbia University Press, NY.

Amundsen, D.C. and Wright, H.E. (1979) Forest Changes in Minnesota at the End of the Pleistocene, *Ecological Monographs*, 49, 1-16.

Anderla, M. *et al*, (1997) *Chaotics: An Agenda for Business and Society in the 21st Century*, Adamantine Press, Twickenham, UK.

Anderson, J.F. and May, R. (1979) Population Biology of Infectious Diseases: Part One, *Nature*, 280, 361-367.

Andreski, S. (1969) Introduction to: *Principles of Sociology* by Herbert Spencer,. Macmillan, London, ix-xxxvi.

Appignanesi, R. *et al* (1995) *Postmodernism for Beginners*, Icon, Cambridge.

Appignanesi, R. and Lawson, H. (1989) *Dismantling the Truth: Reality in a Postmodern World*, St. Martins Press, NY.

Applewhite, P. (1979) Learning in Protozoa, in *Biochemistry and Physiology of Protozoa* (2nd edn) Academic Press, London.

Aristotle (1941) *The Basic Works of Aristotle;* NY Bks, NY.

Armstrong, S.J. and Botzler, C.J., eds, *Environmental Ethics: Divergence and Convergence*, McGraw-Hill, N.Y., 39-48

Aronson, L.R. (1971) Further Studies on Orientation and Jumping Behaviour in the Gorbiid Fish; Bathygobus soporater, *Annals of the New York Academy of Sciences,* 188, 378-392.

Artigiani, R. (1991) Post-Modernism and Social Evolution: An Inquiry, *World Futures*, 30, 149-161.

Ashley, D. (1991) Playing with the Pieces: The Fragmentation of Social Theory, in P. Wexler, ed, *Critical Theory Now*, Falmer Press, London, 70-97.

Assouni, J (1994) *Metaphysical Myths, Mathematical Practice*, Cambridge University Press, Cambridge.

Auerbach, S.I. (1995) Foreword: George M. Van Dyne--A Reminiscence' in B.C. Patten & S.E. Jorgensen, eds, *Complex Ecology*, Prentice-Hall, N.J., xxvii-xxx.

Ayres, R.U. (1994) *Information, Entropy and Progress*, AIP Press, Woodbury, N.Y.

Babich, B.E. *et al.* (1995) On the Idea of Continental and Postmodern Perspectives in the Philosophy of Science, in B.E. Babich *et al*, eds, *Continental and Postmodern Perspectives in the Philosophy of Science*, Avebury, Aldershot, UK, 1-10.

Bache C.M. and S. Grof (2000) *Dark Night, Early Dawn: Steps to a Deep Ecology of Mind*, SUNY Press, Albany.

Barber, T. (1993) *The Human Nature of Birds*, Bookman, Melbourne.

Barnes, B. and Shapin, S. ,eds, (1979) *Natural Order*, Sage, London.

Barnhill, D.L. and Gottlieb, R.S. (2001) *Deep Ecology and World Religions*, SUNY Press, Albany.

Barry, N. (1982) The Tradition of Spontaneous Order, *in Literature of Liberty*, Sept 1982, 45-67.

Barthes, R. (1970) *S/Z* , transl: R. Miller, (1975), Blackwell, Oxford.

Barthes, R. (1981) *Camera Lucida*, Hill and Wang.

Basar, E., ed, (1990*) Chaos in Brain Function*, Springer-Verlag, Berlin

Baudrillard, J. (1983) *Simulacra and Simulations*, transl: P. Foss *et al*, Semiotext(e), NY.

Bauman, Z. (1992*) Intimations of Postmodernity*, Routledge, London.

Bauman, Z. (1993) *Postmodern Ethics*, Blackwell, MA.

Bazzaz, F.A. and T.W. Sipe (1987) Physiological Ecology, Disturbance, and Ecosystem, Recovery, in E.D. Schulze & H. Zwolfer, eds, *Potentials and Limitations of Ecosystem Analysis*, Springer-Verlag, Berlin, 203-227.

Beck, B.B. (1980) Animal Tool Behavior: *The Use and Manufacture of Tools by Animals*, Garland STPM Press, NY.

Benton, T (1993) *Natural Relations: Ecology, Animal Rights and Social Justice*, Verso, London.

Bell, D. (1979) *The Cultural Contradictions of Capitalism*, 2nd edn, Heinemann, London.

Bertalanffy, L.v. (1967) General System Theory, in N.J. Demerath & R.A. Peterson, eds, System, Change and Conflict, Free Press, NY, 115-140.

Bertalanffy, L.v. (1968*) General Systems Theory: Foundations, Development and Applications Braziller*, N.Y.

Best, S. (1991) Chaos and Entropy: Metaphors in Postmodern and Social Theory, *Science as Culture*, 2, 11, 198-226.

Best, S. and Kellner, D. (1991) *Postmodern Theory: Critical Investigations*, MacMillan, London.

Biehl, J. (1994) Ecology and the Modernization of Fascism in the German Ultra-Right, *Society and Nature*, 2, 21-78

Biehl, J. and P. Staudenmaier (1995) *Ecofascism: Lessons from the German Experience*, AK Press, Edinburgh.

Biemann, K. (1978), Search for Organics, in *Viking Mars Expedition*, 1976, MMR PR Dept, CO.

Biggins, D.R. (1976) Biology and Ideology, *Science Education*, 60, 4, 567-578.
Birch, C. (1990) *On Purpose*, NSW University Press, Sydney.
Black, M. (1979) More About Metaphor, in A. Ortony, ed, *Metaphor and Thought*, Cambridge University Press.
Bloor, D. (1976) *Knowledge and Social Imagery*, Routledge and Kegan Paul, London.
Bohm, D. (1988) Postmodern Science and a Postmodern World, in D.R., ed, *The Reenchantment of Science*, SUNY Press, NY, 57-68.
Bohm, D. (1994) Postmodern Science and a Postmodern World, in C. Merchant, ed, *Key Concepts in Critical Theory: Ecology*, Humanities Press, NJ, 342-350.
Boje, D. M. and R.F. Dennehy (1993) *Managing in the Postmodern World: America's Revolution Against Exploitation*. Kendall/Hunt. Dubuque.
Bonner, J.T. (1980) *The Evolution of Culture in Animals*, Princeton University Press, NJ.
Bonsor, J. (1997) Gaiaforming, *Amateur Astronomy & Earth Sciences*, 2, 2, 26-28.
Bookchin, M. (1990) *The Philosophy of Social Ecology: Essays in Dialectical Naturalism*, Black Rose, Montreal.
Bossomaier, T. and Green, D. (1998) *Patterns in the Sand: Computers, Complexity and Life*, Allen and Unwin, Sydney.
Botkin, D. (1990) *Discordant Harmonies: A New Ecology for the 21st Century*, Oxford University Press, Oxford.
Botkin, D.B. and Sobel, M.J. (1975) Stability in Time-Varying Ecosystems, *American Naturalist*, 109, 625-646.
Bottomore, T.B. (1985) *Theories of Modern Capitalism*, Allen and Unwin, London.
Boucher, D.H. *et al* (1982) The Ecology of Mutualism, *Annual Review of Ecology and Systematics*, 13, 315-347.
Bradbury, J.W. (1984) Social Complexity and Cooperative Behaviour in Delphiniids, in T. Schusterman *et al*, eds, *Dolphin Cognition and Behaviour*, 361-372.
Bassingthwaighte, J.B. *et al*, (1994) *Fractal Physiology*, Oxford University Press, Oxford.
Brennan, A. (1986) Ecological Theory and Value in Nature, *Philosophical Inquiry*, 8, 66-95.
Brennan, A (1988) *Thinking About Nature*, University of Georgia Press, Athens.
Briggs, J. and Peat, D. (1991) *Turbulent Mirror: An Illustrated Guide to Chaos Theory and the Science of Wholeness*, Harper and Row, NY.
Brown, P.R.F. (1996) *The Gaia Hypothesis*, Web Site Published by Mountain Man Graphics, Australia.
Bunyard, P. and Goldsmith, E., eds, (1989) *Gaia and Evolution*, Abbey, Bodmin, UK.
Burrows, C.J. (1990) *Processes of Vegetation Change*, Unwin-Hyman, London.
Cahen, H. (1988) Against the Moral Considerability of Ecosystems, *Environmental Ethics*, 10, 195-216.
Calinescu, M. (1987) *Five Faces of Modernity: Modernism, Avant-Garde, Decadence, Kitsch, Postmodernism*, Duke University Press, Durham, NC.
Callicott, J.B. (1989) *In Defence of the Land Ethic: Essays in Environmental Philosophy*, SUNY Press, Albany, NY.
Callicott, J.B. (1995) Animal Liberation: A Triangular Affair, in R. Elliot, ed, *Environmental Ethics*, Oxford University Press, Oxford, 29-59.

Camazine, S. *et al*, eds, (2001) *Self-Organisation in Biological Systems*, Princeton University Press, Princeton.

Cameron, A.M. and Endean, R. (1982) Renewed Population Outbreaks of a Rare and Specialized Carnivore (The Starfish *Acanthaster plancii*) in a Complex High Diversity System (the Great Barrier Reef), *Proceedings of the 4th International Coral Reef Symposium*, 41-49.

Cannon, W. (1932) *The Wisdom of the Body*, WW Norton, N.Y.

Capra, F. (1975) *The Tao of Physics*, Shambala, Boston.

Capra, F. (1982) *The Turning Point*, Simon & Schuster, NY.

Capra, F. (1988) *Uncommon Wisdom*, Simon & Schuster, NY.

Capra, F. (1994) Systems Theory and the New Paradigm, in C. Merchant, ed, *Key Concepts Critical Theory: Ecology*, Humanities Press, NJ, 334-341.

Capra, F. (1996) *The Web of Life*, HarperCollins, NY.

Carney, H.J. (1989) On Competition and the Integration of Population, Community and Ecosystem Studies, *Functional Ecology*, 3, 637-641.

Carson, R. (1962) *Silent Spring*, Houghton-Mifflin, Boston.

Cartwright, T.J. (1991) Planning and Chaos Theory, *APA Journal*, Winter 1991, 44-56.

Casti, J. (1994) *Complexification*, Harper Collins, NY.

Caswell, H. (1982) Life History Theory and the Equilibrium Status of Populations, *American Naturalist*, 120, 317-339.

Cattelino, P.J. *et al* (1979) Predicting the Multiple Pathways of Succession, *Environmental Management*, 3, 41-50.

Cavalli-Sforza, L.L. (1986) Cultural Evolution. *American Zoologist*, 26, 845-855.

Chesson, P.L. and Case, T.J. (1986) Overview: Non-equilibrium Community Theories: Chance, Variability, History and Co-existence, in J. Diamond and T.J. Case, eds, *Community Ecology*, Harper & Row, NY, 229-239.

Christensen, P. (1989) Hobbes and the Physiological Origins of Economic Science, *History of Political Economy*, 21, 689-709.

Christensen, P (1994) Fire, Motion, and Productivity: the Proto-Energetics of Nature and Economy in Francois Quesney, in P. Mirowski, ed, *Natural Images in Economic Thought*, Cambridge University Press, Cambridge, 249-288.

Cilliers, P. (1998) *Complexity and Postmodernism: Understanding Complex Systems*, London: Routledge.

Clark, N. (1988) The Re-Enchantment of Nature? The Politics of New Ageism and Deep Ecology, in *Ecopolitics III*: Proc. of the 3rd Ecopolitics Conference, Hamilton, NZ.

Clarke, A.C. (1996) *The Snows of Olympus*, Victor Gollancz, London.

Clayton, A. (1993) Systems Theory: Some Caveats, *Environmental Values*, 2, 159-161.

Clements, F.E. (1905) *Research Methods in Ecology*. (Repr. 1977) University Printing Co, Lincoln, NE.

Clements, F.E. (1916) *Plant Succession: An Analysis of the Development of Vegetation*, Carnegie Institution, Washington, D.C.

Clements, F.E. (1936) Nature and Structure of the Climax, *Journal of Ecology*, 24, 252-284.

Cobb, J.B. (1988) Ecology, Science and Religion: Toward a Postmodern Worldview, in D.R. Griffin, ed, *The Reenchantment of Science*, SUNY Press, NY, 99-122.

Cohen, I. (1994) *Interactions: Some Contact Between The Natural and Social Sciences*, MIT Press, MA.

Cohen, J. and Stewart, I. (1994) *The Collapse of Chaos*, Viking, London.
Connell, J.H. and Slatyer, R.O. (1977) Mechanisms of Succession in Communities and Their Role in Stability and Organisation, *American Naturalist*, 111, 1119-1149.
Connell, J.H. and Sousa, W.P. (1983) On the Evidence Needed to Judge Ecological Stability and Persistence, *American Naturalist*, 121, 789-824.
Conway, G. (1976) Man Versus Pests, in R. May, ed, *Theoretical Ecology: Principles and Applications*, Blackwell, Oxford.
Crawley, M.J. (1983) *Herbivory: The Dynamics of Animal-Plant Interactions*, Blackwell, Oxford.
Cromartie, M., ed, (1995) *The 9 Lives of Population Control*, Eerdmans, Grand Rapids.
Davidson, M. (1983) *Uncommon Sense: The Life and Thought of Ludwig von Bertalanffy*, Tarcher, LA.
Davies, P. (1983) *God and the New Physics*, Simon and Schuster, NY
Davies, P (1987) *The Cosmic Blueprint*, Penguin, Harmondsworth.
Davies, P. (1990) *Chaos Frees the Universe*, New Scientist, 128 (6th Oct. 1996), 36-39.
Davies, P. (1993) *The Mind of God*, Penguin, London
Davies, P (1994) A Vision of Science in the 21st Century, *21st Century*, 12, 102-103.
Davies, P (1996) The Future of God, *Sydney Morning Herald*, Dec 21st, 6.
Davies, P. (1996) Basic Urges Take Some Stopping, *The Australian*: HEC Column, Cosmics: The Australian Online--14/8/96.
Davies, P. (1998) Ants in the Machine, *Sydney Morning Herald*, Oct 17th, 6s.
Davis, M.B. (1981) Quaternary History and the Stability of Forest Communities, in D. West *et al*, eds, *Forest Succession: Concepts and Applications*, Springer-Verlag, Berlin.
Dawkins, R (1982) *The Extended Phenotype*, Freeman, San Francisco.
DeAngelis, D.L. (1995) The Nature and Significance of Feedback in Ecosystems, in B.C. Patten and S.E. Jorgensen, eds, *Complex Ecology*, Prentice Hall, NJ, 450-467.
Delcourt, H.R. (1979) Late Quarternary Vegetation History of the Eastern Highland Rim and Adjacent Cumberland Plateau of Tennessee, *Ecological Monographs*, 49, 218-237.
Delcourt, H.R. et al (1982) Dynamic Plant Ecology: The Spectrum of Vegetational Change in Space and Time, *Quaternary Science Review*, 1, 153-176.
Deleuze, J and Guattari, F. (1987) *A Thousand Plateaus*, Minnesota University Press, Minneapolis.
Demerath, N.J. (1967) Synechdoche and Structural Functionalism, in N.J. Demerath and R.A. Peterson, eds, *System, Change and Conflict*, Free Press, NY.
Dendrinos, D.S. (1990) *Chaos and Socio-spatial Dynamics*, Springer-Verlag, NY.
Derrida, J. (1974) White Mythology, *New Literary History*, 6, 1, 4-74.
Derrida, J. (1976) *Of Grammatology*, transl: G. Spivak, Johns Hopkins University Press, MD.
Devall, B. and Sessions, G. (1985) *Deep Ecology: Living As if the Earth Really Mattered*, Gibbs M. Smith, Layton.
Dickens, P. (1992) *Society and Nature: Towards a Green Social Theory*, Harvester Wheatsheaf, London.
Digregario, B.E. *et al* (1997) *Mars: The Living Planet*, Frog Limited.
DiZeriga, G. (1993), Unexpected Harmonies: Self-Organisation in Liberal Modernity and Ecology, *The Trumpeter*, 10, 25-32.

Docherty, T., ed, (1993) *Postmodernism: A Reader*, Harvester-Wheatsheaf, Hemel Hempstead.
Donovan, S.K. (1989) *Mass Extinctions: Processes and Evidence*, Belhaven Press, London.
Doolittle, W.F. (1981) Is Nature Really Motherly? *Coevolutionary Quarterly*, 19, 58-63.
Drake, J.A. (1990) Communities as Assembled Structures: Do Rules Govern Pattern? *Trends in Ecology and Evolution*, 5, 159-64.
Drury, W.H. and Nisbet, I.C. (1973) Succession, *Journal of Arnold Arboretum*, 54, 331-368.
Dryzak, J. (1987) *Rational Ecology: Environment and Political Economy*, Blackwell, Oxford.
Dunlap, R.E. and Morrison, D.E. (1986) Environmentalism and Elitism: a Conceptual and Empirical Analysis. *Environmental Management*, 10, 581-589.
Dupre, J. (1993) *Disorder of Things: Metaphysical Foundations of the Disunity of Science*, Harvard University Press, Cambridge, MA.
Durning, A. (1997) *Misplaced Blame: the Real Roots of Population Growth*, Seattle, Northwest Environment Watch.
Dyke, C. (1988) *The Evolutionary Dynamics of Complex Systems: A Study in Biosocial Complexity*, Oxford University Press, N.Y.
Eagleton, T. (1986) *Against the Grain*, Verso, London.
Easthope, A. (1998) Postmodernism and Critical and Cultural Theory, in S. Sim, ed, *Postmodern Thought*, Icon Books, Cambridge, 15-27.
Eckersley, R. (1989) Green Politics and the New Class: Selfishness or Virtue? *Political Studies*, 40, 314-332.
Edson, M.M. *et al* (1981) Emergent Properties and Ecological Research, *American Naturalist*, 118, 593-596.
Egerton, F.N. (1973) Changing Concepts in the Balance of Nature, *Quarterly Review of Biology*, 48, 322-350.
Elliot, R., ed, (1995) *Environmental Ethics*, Oxford University Press, Oxford
Engleberg, J. and Boyarsky, L.L. (1979) The Non-Cybernetic Nature of Ecosystems, *American Naturalist*, 114, 317-24.
Fabel, A.J. (1994) Environmental Ethics and the Question of Cosmic Purpose, *Environmental Ethics*, 16, 303-314.
Faegri, K. and van der Pijl, L. (1971) *The Principles of Pollination Ecology*, Pergamon Press, Oxford.
Farganis, J. (1993) *Readings in Social Theory: The Classic Tradition to Postmodernism*, McGraw-Hill, NY.
Feenberg, A. (1979) Beyond the Politics of Survival, *Theory and Society*, 7, 319-361.
Fenchel, T. and Christensen, F.B. (1976) *Theories of Biological Communities*, Springer-Verlag, NY.
Ferre, F. (1988) Religious World Modeling and Postmodern, in D.R. Griffin, ed, *The Reenchantment of Science*, SUNY Press, NY, 87-98.
Ferre, F (1989) Obstacles on the Path to Organismic Ethics: Some Second Thoughts, *Environmental Ethics*, 11, 231-241.
Ferre, F. (1993) *Hellfire and Lightning Rods: Liberating Science, Technology and Religion*, Orbis Books, NY.
Ferre, F (1996) *Being and Value: Toward a Constructive Postmodern Metaphysical*, SUNY Press, Albany.
Feyerabend, P. (1993) *Farewell to Reason*, Verso, London.

Foucault, M. (1974) *The Archeology of Knowledge*, Tavistock, London

Foucault, M. (1983) 'Preface' to G. Deleuze & F. Guattari, *Anti-Oedipus*, University of Minnesota Press, Minneapolis.

Foucault, M. (1986) What is Enlightenment?, in P. Rabinow, ed, *The Foucault Reader*, Penguin, Harmondsworth.

Foucault, M. (1994) Geneology and Social Criticism, in S. Seidman, ed, *The Postmodern Turn: New Perspectives on Social Theory*, Cambridge University Press, Cambridge, 39-45.

Frank, P.W. (1968) Life Histories and Community Stability, *Ecology*, 49, 355-357.

Fraser, N., and L. Nicholson (1994) Social Criticism Without Philosophy: an Encounter between Feminism and Postmodernism, in S. Seidman, ed, *The Postmodern Turn: New Perspectives on Social Theory*, Cambridge University Press, Cambridge, 242-261.

Gare, A. (1995) *Postmodernism and the Environmental Crisis*, Routledge, London.

Geyer, F. (1994) The Challenge of Sociocybernetics, *Paper Presented to the World Congress of Sociology*, Bielefeld, July 18-24, 1994.

Glacken, C.J. (1967) *Traces on the Rhodian Shore: Nature and Culture in Western Thought From Ancient Times to the End of the 18th Century*, University of California Press, Berkeley.

Gleason, H.A. (1926) The Individualistic Concept of the Plant Association, *Bulletin of the Torrey Botanical Club*, 53, 1-20.

Gleick, J. (1987) *Chaos*, Penguin, London.

Goener, S.J. (1993) Reconciling Physics and the Order-Producing Universe: Evolutionary Competence and the New Vision of the Second Law, *World Futures*, 36, 167-179.

Goldsmith, D. (1998) *The Hunt for Life on Mars*, Plume, NY.

Goldsmith, E. (1993) *The Way: An Ecological Worldview*, Shambhala, N.Y.

Goldstein, S.F. (1992) Flagellar Beat Patterns in Algae, in M. Melkonian, ed, *Algal Cell Motility*, Chapman Hall, NY, 99-145.

Golley, F.B. (1993) *A History of the Ecosystem Concept in Ecology*, Yale University Press, New Haven.

Gould, S.J. (1988) *Time's Arrow, Time's Cycle: Myth and Metaphor in the Discovery of Geological Time*, Penguin, Harmondsworth.

Goulden, C.E. (1977) *Changing Scenes in the Life Sciences, 1776-1976,* SP12, Academy of Natural Sciences, Philadelphia.

Gowdy, J.M. (1994) Progress and Environmental Sustainability, *Environmental Ethics*, 16, pp41-55

Gowdy, J.M. (1994) Further Problems with Neo-Classical Environmental Economics, *Environmental Ethics*, 16, 161-171.

Grant, I.H. (1998) Postmodernism and Science and Technology, in S. Sim, ed, *Postmodern Thought*, 65-77

Greene, J.C. (1981) *Science, Ideology and Worldview: Essays in the History of Evolution Ideas*, University of California Press, Berkeley.

Grenz, S. (1996) *A Primer on Postmodernism*, Eerdmans, Grand Rapids.

Griffin, D.R. (1988) Introduction to the SUNY Series in Constructive Postmodern Thought, in D.R. Griffin, ed, *The Reenchantment of Science*, SUNY Press, NY, ix-xii.

Griffin, D.R., ed, (1988) *The Reenchantment of Science*, SUNY Press, NY.

Haber, H.F. (1994) *Beyond Postmodern Politics*, Routledge, NY.

Habermas, J, (1987) *The Philosophical Discourse of Modernity*, transl: T. McCarthy, Beacon Press, Boston.

Hagen, J. (1992) *An Entangled Bank: The Origins of the Ecosystem Concept*, Rutgers University Press, New Brunswick.

Hall, S., ed, (1992) *Understanding Modern Societies* (Vol. 1-4) , Polity Press, Cambridge.

Hall, G.M. (1997) *The Ingenious Mind of Nature: Deciphering the Patterns of Man, Society, and the Universe*, Plenum Trade, NY.

Hallman, H.O. (1991) Nietzsche's Environmental Ethics, *Environmental Ethics*, 13, 99-125.

Haney, K.M. (1994) *Intersubjectivity Revisited: Phenomenology and the Other*, Ohio University Press, Athens.

Harris, G.P. (1986) *Phytoplankton Ecology: Structure, Function and Fluctuation*, Chapman and Hall, London.

Harvey, D. (1990) *The Condition of Postmodernity*, Blackwell, Oxford.

Hassan, I. (1987) *The Postmodern Turn: Essays in Postmodern Theory and Culture*, Ohio State University Press, Ohio.

Hayek, F.A. (1967) *Studies in Philosophy, Politics and Economics*, Chicago University Press, Chicago.

Hayles, N.K. (1990) *Chaos Bound: Orderly Disorder in Contemporary Literature and Science*, Cornell University Press, Ithaca, N.Y.

Hayles, N.K., ed, (1991*) Chaos and Order: Complex Dynamics in Literature and Science*, Chicago University Press, Chicago.

Heath, S. (1998) Intertextuality, in M. Payne, ed, *A Dictionary of Cultural and Critical Theory*, Blackwell, Oxford.

Heims, S.J. (1991) *The Cybernetics Group*, MIT Press, Cambridge, MA.

Heinselman, J and Wright, I. (1973*)* The Ecological Role of Fire in Natural Conifer Forests of Western and Northern North America, *Quaternary Research*, 3, 317-513.

Heller, A. and F. Feher. (1988) *The Postmodern Political Condition*, Polity Press, Cambridge.

Heyl, S. (1997) The Harvard Pareto Circle. Reproduced on the *Science as Culture* website.

Hollinger, R. (1994*) Postmodernism and the Social Sciences: A Thematic Approach*, Sage Publ., California.

Holm, C.R., eds, (1995) *Population: Opposing Viewpoints*, Greenhaven, San Diego.

Horigan, S (1988) *Nature and Culture in Western Discourses*, Routledge, London.

Hutter, M. (1994) Organism as a Metaphor in German Economic Thought, in P. Mirowski, ed, *Natural Images in Economic Thought*, Cambridge University Press, Cambridge, 289-321.

Jackson, R.R. (1995) What is that spider thinking? *New Zealand Science Monthly*, February: 6-8.

Jackson, R.R. & Pollard, S.D. 1996. *Predatory behavior of jumping spiders. Annual. Review of Entomology,* 41: 287-308

Jagtenberg, T. and McKie, S. (1997*) Eco-Impacts and the Greening of Postmodernity*, Sage, CA.

Jameson, F. (1991*) Postmodernism, or, the Cultural Logic of Late Capitalism*, Verso, London.

Jantsch, E. (1980), *The Self-Organising Universe*, Pergamon Press, Oxford.

Jencks, C. (1986) *What is Postmodernism?*, Academy Edition, London.

Joseph, L.E. (1990) *Gaia: The Growth of an Idea*, Arkana, NY.

Kauffman, S (1993) *The Origins of Order*, Oxford University Press, NY.

Kauffman, S. (1995) *At Home in the Universe*, Oxford University Press, NY.

Kawai, M. (1965) Newly Acquired Pre-Cultural Behaviour of the Natural Troop of Monkeys on Koshua Islet, *Primates*, 66, 21-36.

Kember, S. (1996) Feminist Figuration and the Question of Origin, in G. Robertson *et al*, eds, *FutureNatural*, Routledge, London, 256-269.

Kempton, W. *et al* (1995) *Environmental Values in American Culture*, MIT Press, MA

Keolb, C., ed, (1996) *Nietzsche as Postmodernist: Essays Pro and Contra*, SUNY Press, Albany.

Kingsland, S.E. (1985) *Modeling Nature*, Chicago University Press, Chicago.

Kingsland, S.E. (1994) Economics and Evolution: Alfred James Lotka and the Economy of Nature, in P. Mirowski, ed, *Natural Images in Economic Thought*, Cambridge University Press, Cambridge, 231-248.

Klamer, A. and Leonard, T.C. (1994) So What's an Economic Metaphor?, P. Mirowski, ed, *Natural Images in Economic Thought*, Cambridge University Press, Cambridge, 20-54.

Kley, R. (1994) *Hayek's Social and Political Thought*, Clarendon, Oxford.

Krohn, W. *et al* (1990) Self-Organisation--The Convergence of Ideas: An Introduction, in W. Krohn *et al*, eds, *Self-Organisation: Portrait of a Scientific Revolution*, Kluwer Academic, Dordrecht.

Krippner, S. (1991) The Holistic Paradigm, *World Futures*, 30, 133-140.

Kristeva, J. (1967) Word, Dialogue and Novel, in T. Moi, ed, *The Kristeva Reader Blackwell*, Oxford, 39-45.

Kroker, A., and Cook, D. (1991*) The Postmodern Scene: Excremental Culture and Hyper-Aesthetics*, 2nd edn, MacMillan, London

Kropotkin, P.A. (1902) *Mutual Aid: A Factor in Evolution*, Wilheinium, London.

Kuppers, B-O. (1990) On a Fundamental Shift in the Natural Sciences, in W. Krohn *et al*, eds, *Self-Organisation: Portrait of a Scientific Revolution*, Kluwer Academic, Dordrecht, 51-63.

Kwa, C.L. (1987) Representations of Nature Mediating Between Ecology and Science Policy: The Case of the International Biological Programme, *Social Studies of Science*, 17, 413-442.

Kwa, C.L. (1993) Radiation Ecology, Systems Ecology, and the Management of the Environment in M. Shortland, ed, *Science and Nature*, (BSHS Monograph) 213-249.

Lacan, J. (1982) The Meaning of the Phallus, in J. Mitchell and J. Rise, eds, *Feminine Sexuality: Jacque Lacan and the Ecole Freudienne*, Tavistock. London, 146-178 (orig. publ. 1958).

Lakoff, G. and Johnson, M. (1980) *Metaphors We Live By*, University of Chicago Press, Chicago.

Lash, S. and Urry, J. (1987) *The End of Organised Capitalism*, Polity, London.

Laszlo, E. (1972) *Introduction to Systems Philosophy: Toward a New Paradigm of Contemporary Thought*, Jordon and Breach, NY.

Laszlo, E. (1997) "Planetary Consciousness: Our Next Evolutionary Step" *Cybernetics and Human Knowing: A Journal of Second Order Cybernetics & Cyber-Semiotics*, 4, No.4 (1997).

Latour, B and Woolgar, S. (1979*) Laboratory Life: The Construction of Scientific Facts*, Princeton University Press, Princeton.

Latour, B. *Science in Action: How to Follow Scientists and Engineers Through Society* (Cambridge: Harvard University Press; 1988).

Lauwerier, H. (1991) *Fractals: Images of Chaos*, Penguin, Harmondsworth.

Layder, D. (1994) *Understanding Social Theory*, Sage, London.

Lemert, C. (1992) General Social Theory, Irony, Postmodernism, in S. Seidman and D.G. Wagner, eds, *Postmodernism and Social Theory*, Blackwell, Cambridge, MA, 17-46.

Lemkow, A. (1995) *The Wholeness Principle: Dynamics of Unity Within Science, Religion, and Society*, Quest Books, Wheaton, Revised edn.

Lesourne, J. (1992) *The Economics of Order and Disorder: The Market as Organizer and Creator*, Clarendon, Oxford.

Levins, R. and Lewontin, R. (1985) *The Dialectical Biologist*, Harvard University Press, Cambridge.

Levin, S.A. (1999) *Fragile Dominion: Complexity and the Commons*, Helix Books.

Lewin, R. (1996) All for One, One for All, *New Scientist*, 152 (14 Dec 1996), 28-33.

Lewis, B. (1998) Postmodernism and Literature, in S. Sims, ed, *Postmodern Thought*, Icon Books, Cambridge, 121-132.

Lewontin, R.C. *et al* (1984) *Not In Our genes: Biology, Ideology and Human Nature*, Pantheon Books.

Lilienfeld, R (1988) *The Rise of Systems Theory: An Ideological Analysis*, Krieger (orig. publ. 1978).

Limoges, C. and Menard, C. (1994), Organisation and the Division of Labour: Biological Metaphors at Work in Alfred Marshall's 'Principles of Economics', in P. Mirowski, ed, *Natural Images in Economic Thought*, Cambridge University Press, Cambridge.

Lindauer, M. (1961) Communication Among Social Bees, Harvard University Press, Cambridge, MA.

Lindeman, R.L. (1941) The Developmental History of Cedar Creek Bog, , *American Midland Naturalist*, 25, 101-112.

Lindeman, R.L. (1942) The Trophic-Dynamic Aspect of Ecology, *Ecology*, 23, 399-418.

List, P. (1994) *Radical Environmentalism: Philosophy and Tactics*, Wadsworth, California.

Lovelock, J. (1979) *Gaia: A New Look at Life*, Oxford University Press, Oxford.

Lovelock, J. (1987) A Model for Planetary and Cellular Dynamics, in W.I. Thompson, ed, *Gaia: A Way of Knowing*, Lindisfarne Press, Lindisfarne, MA.

Lovelock, J. (1988) *The Ages of Gaia*, WW. Norton & Co, NY.

Lovelock, J. (1991) *Healing Gaia: Practical Medicine for the Planet*, Harmony, NY.

Lovelock J., (1992) A Numerical Model for Biodiversity, *Philosophical Transactions of the Royal Society of London*, B338, 383-391.

Lovelock, J. (1994) Gaia, in C. Merchant, ed, *Key Concepts in Critical theory: Ecology*, Humanities Press, NJ, 351-359.

Lovelock, J. and Allaby, M. (1985*) The Greening of Mars*, Warner Books, NY.

Loy, D.R. (1997) The Religion of the Market, *Journal of American Academy of Religion*, 65, 275-290.

Lucie-Smith, E. (1992) *Movements in Art Since 1945*, Thames and Hudson, London.
Lyotard, J-F. (1984) *The Postmodern Condition*, transl: by G. Bennington and B. Massumi Manchester University Press, Manchester (orig. publ. 1979).
Lyotard, J-F. (1988) *The Differend: Phrases in Dispute*, transl: by G. Van Der Abbele, Manchester University Press, Manchester (orig. publ. 1983).
Macy, J. (1994) Toward a Healing of Self and World, in C. Merchant, ed, *Key Concepts in Critical Theory: Ecology*, Humanities Press, NJ, 292-298.
Mainzer, K. (1995) *Thinking in Complexity: The Complex Dynamics of Matter, Mind, and Mankind*, Springer-Verlag, Berlin.
Maley, B. (1994) *Ethics and Ecosystems*, Centre for Independent Studies, Sydney.
Mandelbrot, B.B. (1982) *The Fractal Geometry of Nature*, W.H. Freeman.
Margalef, R. (1968) *Perspectives in Ecological Theory*, Chicago University Press, Chicago.
Margulis, L. and G. Hinkle. (1991) The Biota of Gaia: 150 Years of Support for the Environmental Science, in S.H. Schneider and P.J. Boston, eds, *Scientists on Gaia*, MIT Press, Cambridge, MA, 11-18.
Margulis, L. and D. Sagan, (1992) Can Mars be Colonized?, *Paper Presented to the 29th Plenary Meeting of the International Astronautical Federation*, Washington, D.C.
Margulis, L. and O. West, (1993) Gaia and the Colonization of Mars, *GSA Today*, 3(11), 277-291.
Margulis, L. and D. Sagan (1995) *What is Life?* Simon & Schuster, NY.
Marietta, D.E. (1993) Environmental Holism and Individuals, in S.J. Armstrong and R.G. Botzler, eds, *Environmental Ethics: Divergence and Convergence*, McGraw-Hill, N.Y.
Marshall, A (1993) *Ethics and the Extraterrestrial Environment*, Journal of Applied Philosophy, 10, 227-237.
Marshall, A (1995) Development and Imperialism in Space, *Space Policy*, 11, 1, 241-252.
Marshall, A. (1997) Gaian Ecology and Environmentalism, *Wild Earth*, 7, 76-81.
Marshall, A. (1997) Another Green World?, *Quest*, 1, 3, 38-51.
Marshall, A (1998) A Postmodern Natural History of the World: Eviscerating the GUTs from Ecology and Environmentalism, *Studies in the History and Philosophy of Science*: Part C, 29, 137-164.
Martin, B. (1978) The Selective Usefulness of Game Theory, *Social Studies of Science*, 8, 85-110.
Martin-Smith, M. (1997) Microcosm and Macrocosm, *Quest*, 1, 3, 31-32.
Matthews, F. (1991) *The Ecological Self*, Routledge, London.
Maturana, H. and Varela, F. (1980) *Autopoiesis and Cognition*, D. Reidel Publishing Company.
May, R. (1986) The Search for Patterns to the Balance of Nature: Advances and Retreats, *Ecology*, 67, 1115-1126.
May, R. and Anderson, R.M. (1979) Population Biology of Infectious Diseases: Part Two, *Nature*, 280, 455-461.
Mayr, O. (1986) *Authority, Liberty and Automatic Machinery in Early Modern Europe*, Johns Hopkins University Press, Baltimore.
McClosky, D. (1985) *The Rhetoric of Economics*, University of Wisconsin Press, Madison.

McGrew, W.C. *et al*, (1979) Chimpanzees, Tools and Termites: Cross Cultural Comparisons of Senegal and Rio Mini, *Man*, 14, 185-214.

McIntosh, R.P. (1985) *The Background to Ecology*, Cambridge University Press, NY.

McLaughlin, A. (1990) Ecology, Capitalism and Socialism, *Socialism and Democracy*, 10, 69-102.

Merchant, C. (1980) *The Death of Nature: Women, Ecology and the Scientific Revolution*, Harper & Row, San Francisco.

Merchant, C., ed, (1994) *Key Concepts in Critical Theory: Ecology*, Humanities Press, NJ.

Merry, U. (1995) *Coping With Uncertainty: Insights from the New Sciences of Chaos, Self-Organisation and Complexity*, Praeger, Westport, CT.

Michener, C.D. (1974) *The Social Behavior of Bees*, Harvard University Press, Cambridge, MA.

Miller, D. (1982) Metaphor, Thinking and Thought, *ETC.: A Review of General Semantics*, 39, 134-50.

Miller, D. (1982) Metaphor, Thinking and Thought: Part Two, *ETC.: A Review of General Semantics*, 39, 242-256.

Miller, D. (1983) Metaphor and Culture, *Paper Presented to the 1983 Culture Seminar*, University of Melbourne, Australia.

Milovanovic, D. (1997) *Postmodern Criminology*, Garland, New York.

Mingers, J. (1995) *Self-Producing Systems*, Plenum, NY.

Mirowski, P. (1988) *Against Mechanism: Protecting Economics from Science*, Rowman & Littlefield, NJ.

Mirowski, P. (1989) *More Heat Than Light: Economics as Social Physics, Physics as Nature's Economics*, Cambridge University Press, Cambridge.

Mirowski, P. (1994) Doing What Comes Naturally: Four Metanarratives on What Metaphors are For, in P. Mirowski, ed, *Natural Images in Economic Thought: Markets Read in Tooth and Claw*, Cambridge University Press.

Mirowski, P., ed, (1994), *Natural Images in Economic Thought: Markets Read in Tooth and Claw*, Cambridge University Press, Cambridge.

Moore, R.D. (1992) From Science to Mythology: A New Vision of Reality, in J. Kliest and B.A. Butterfield, eds, *Mythology: From Ancient to Postmodern*, Lang, NY.

More, M. (1991) Order Without Orders, *Extropy*, 3, 21-32.

Morton, A.G. (1981) *History of Botanical Science*, Academic Press, NY.

Muller, J.Z. (1993) *Adam Smith in His Time and Ours: Designing the Decent Society*, Free Press, NY.

Myers, N. (1985) *The Gaia Atlas of Planet Management*, Pan Books, U.K.

Nabhan, G.P. (1995) Cultural Parallax in Viewing North American Habitats, in M.E. Soule and G. Lease, eds, *Reinventing Nature: Responses to Postmodern Deconstruction*, Island, Washington, D.C.

Naess, A. (1973) The Shallow and The Deep, Long-Range Ecology Movement, *Inquiry*, 16, 95-100.

Naess, A. (1986) The Deep Ecological Movement: Some Philosophical Aspects, *Philosophical Inquiry*, 8, 10-31.

Naess, A. (1989) *Ecology, Community, Lifestyle*, Cambridge University Press, Cambridge, U.K.

Naess, A. (1994) Deep Ecology, in C. Merchant, ed, *Key Concepts in Critical Theory: Ecology*, Humanities Press, NJ, 120-124.

Neville, R. (1998) The Future Isn't What It Used To Be, *Good Weekend* May 16th, 1998, 16-22.

Nicholson, S. (1992) The Living Cosmos, in S. Nicholson and B. Rosen (Eds*) Gaia's Hidden Life*, Quest Books, Wheaton, Ill.

Nietzsche, F. (1979) Philosophy and Truth, in D. Breazeal, ed, *Selections from Nietzsche's Notebooks of the Early 1870s*, Humanities Press, NJ.

Nordenskiold, E. (1928) *The History of Biology: A Survey*, A. Knopf, NY.

Odum, E.P. (1971) *Fundamentals of Ecology*, 3rd edn (1st edn 1953), Saunders, Philadelphia.

O'Neill, R.V. and Reichle, D.E. (1980) Dimensions of Ecosystem Theory, in *Forests: Fresh Perspectives from Ecosystem Analysis*, 40th Annual Biology Colloquium, Oregon State University Press, Corvallis.

O'Neill, R.V. *et al* (1986) *A Hierarchical Concept of the Ecosystem*, Princeton University Press, NJ.

Pace, D. (1983) *Claude Levi-Strauss: Structuralism and Sociological Theory*, Routledge and Kegan Paul, Boston.

Paine, R.T. (1980) Food Webs: Linkage, Interaction, Strength and the Community Infrastructure, *Journal of Ecology*, 2, 159-170.

Parsons, T. (1967) *The Social System*, 2nd Edn (1st Edn 1951), Routledge, London.

Parker, D. and Stacey, R. (1995*) Chaos, Management and Economics*, Centre for Independent Studies, St. Leonards, NSW.

Partridge, E. (1984) Nature as a Moral Resource, *Environmental Ethics*, 6, 102-130.

Peat, F.D. (1991*) The Philosopher's Stone: Chaos, Synchronicity and the Hidden Order of the World*, Bantam Books, NY.

Peitgen, H-O and Saupe, S., eds, (1988) *The Science of Fractal Images*, Springer-Verlag, NY.

Pepper, D. (1984) *The Roots of Modern Environmentalism*, Croon Helm, Beckenham, UK.

Pepper, D. (1993) *Ecosocialism: From Deep Ecology to Social Justice*, Routledge, London.

Pepperberg, I.M. (1983) Cognition in the African Grey Parrot: Preliminary Evidence for Auditory/Vocal Comprehension of the Class Concept, *Animal Learning and Behaviour*, 11, 179-185.

Pepperberg, I.M. (1991) A Communicative Approach to Animal Cognition: A Study of the Conceptual Abilities of the Grey Parrot, in C.A. Ristau, ed, *Cognitive Ethology: The Minds of Other Animals*, Lawrence Erlbaum, NJ, 153-186.

Peterkin, G.F. and C.R. Tubbs (1965) Woodland Regeneration in the New Forest, Hampshire Since 1650, *Journal of Ecology*, 2, 159-170.

Peters, T. (1987) *Thriving on Chaos*, Knopf, N.Y.

Pfeil, F. (1988) Postmodernism as a Structure of Feeling, in L. Grossberg *et al*, eds, *Cultural Studies*, Routledge, NY, 592-599.

Pickett, S.T.A. and White, P.S., eds, (1985*) The Ecology of Natural Disturbance and Patch Dynamics*, Academic Press, Orlando, Fla.

Porritt, J. (1984) *Seeing Green*, Blackwell, Oxford.

Porter, T. (1990) Natural Science and Social theory, in R.C. Olby *et al* ,eds, *Companion to the History of Modern Science*, Routledge, London, 1024-1043.

Poster, M. (1990) *The Mode of Information: Poststructuralism and Social Context*, Chicago University Press, Chicago.

Prigogine, I. and Stengers, I. (1984*) Order Out of Chaos*, Bantom Books, N.Y.

Probst, G.J.B. *et al* (1994) *Self-Organisation and Management of Social Systems: Insights, Promises, Doubts and Questions*, Springer, NY.

Pryor, K. and Norris, K.S., eds, (1991) *Dolphin Societies and Puzzles*, University of California Press, 349-363.

Rasch, W. *et al*, eds (2000) *Observing Complexity: Systems Theory and Postmodernity*, University of Minnesota Press, Minneapolis.

Real, L.A. and Brown, J.A. (1991*) Foundations of Ecology: Classic Papers with Commentaries*, University of Chicago Press, Chicago.

Redclift, M. (1987) *Sustainable Development: Exploring the Contradictions*, Methuan, London.

Regan, T (1983) *The Case for Animal Rights*, University of California Press, Berkeley.

Richardson, G.P. (1992) *Feedback Thought in Social Science and Systems Theory*, University of Pennsylvannia Press, Philadelphia.

Richardson, J. (1980) The Organismic Community: Resilience of an Embattled Concept, *BioScience*, 30, 465-471.

Richmond, R.C. *et al* (1975) A Search for Emergent Competitive Phenomena: The Dynamics of Multispecies Drosophila Systems, *Ecology*, 56, 709-714.

Ricouer, P. (1977) *The Rule of Metaphor*, University of Toronto Press, Toronto.

Riffaterre, M. (1983) *Text Production*, Columbia University Press, NY.

Ristau, C.E. ed, *Cognitive Ethology: The Minds of Other Animals*, Lawrence Erlbaum, NJ

Robertson, D. (1993) *Penguin Dictionary of Politics*, (New edn) Penguin, London.

Rollo, C.D. (1995) *Phenotypes: Their Epigenetics, Ecology and Evolution*, Chapman and Hall, London.

Rorty, R. (1980) *Philosophy and the Mirror of Nature*, Blackwell, Oxford.

Rosen, R. (1972) Review of Trends in General Systems Theory, *Science*, 177, 508-509.

Rosenau, P. (1992) *Postmodernism and the Social Sciences*, Princeton, Princeton University Press.

Roszak, T. (1979) *Person/Planet: the Creative Disintegration of Industrial Society*, Victor Gollancz, London.

Roszak, T. (1991) *The Voice of the Earth*, Simon and Schuster, NY.

Roth, G and H. Schwegler (1990) Self-Organisation, Emergent Properties and the Unity of the World, in W. Krohn *et al*, eds, *Self-Organisation: Portrait of a Scientific Revolution*, Kluwer Academic, Dordrecht, 36-50.

Rothschild, M. (1990) *Bionomics: The Inevitability of Capitalism*, Henry Holt.

Rowe, S. (1997) "From Reductionism to Holism in Ecology and Deep Ecology", *The Ecologist* , 27, 4, 147-151.

Russell, R.J. *et al* (1990*), Chaos and Complexity: Scientific Perspectives on Divine Action*, Specola Vationa, Vatican City.

Sagan, C (1981) *Cosmos*, MacMillan, London.

Sagan, D. and Margulis, L. (1988) Gaia and Biospheres, in P. Bunyard and E. Goldsmith, eds, *Gaia: the Thesis, the Mechanisms, the Implications*, Quintrell, Wadebridge, UK.

Sahlins, M. (1977*) The Use and Abuse of Biology: An Anthropological Critique of Sociobiology*, Tavistock, London.

Sahtouris, E. (1992) The Dance of Life, in S. Nicholson & B. Rosen, eds, *Gaia's Hidden Life*, Quest Books, Wheaton, Ill.

Salt, G.W. (1979) A Comment on the Use of the Term 'Emergent Properties', *American Naturalist*, 113, 145-148.

Sassower, R. (1995) *Cultural Collisions: Postmodern Technoscience*, Routledge, London.

Schlain, L. (1991) *Art and Physics: Parallel Visions in Space, Time and Light*, Morrow, NY.

Schulze, E.D. and Zwolfer, H., eds, (1987) *Potentials and Limitations of Ecosystem Analysis*, Springer Verlag, Berlin.

Schweber, S. (1980) Darwin and the Political Economists, *Journal of the History of Biology*, 13, 195-289.

Schweickart, R. (1987) Gaia, Evolution, and the Significance of Gaia, *IS Journal*, 2, 2, 29.

Scott, D. (1998) Postmodernism and Music, in S. Sims, ed, *Postmodern Thought*, 134-146.

Scott, G. (1991) *Time, Rhythm and Chaos In the New Dialogue with Nature*, Iowa State University Press, Ames.

Sessions, G. (1994) Ecocentrism and the Anthropocentric Detour, in C. Merchant, ed, *Key Concepts in Critical Theory: Ecology*, Humanities Press, NJ, 125-139.

R. Sheldrake *et al,* (2001) *Chaos, Creativity and Cosmic Consciousness*, Park Street Press.

Shugart, H.H. (1984) *A Theory of Forest Dynamics: The Ecological Implications of Forest Succession Models*, Springer-Verlag, NY.

Shugart, H.H. and O'Neill. R.V., eds, (1979) *Systems Ecology*, Dowden Hutchison and Ross, Stroudsberg, PA.

Sim, S., ed, (1998) *Postmodern Thought*, Icon Books Cambridge, U.K.

Sim, S. (1998) Postmodernism and Philosophy, in S. Sim, ed, *Postmodern Thought*, Icon, Cambridge, 3-14.

Simmons, I.G. (1993) *Interpreting Nature: Cultural Constructions of the Environment*, Routledge, London.

Skolomowski, H. (1981) *Eco-philosophy: Designing New Tactics for Living*, Boyars, Boston.

Slack, J.O. and L.A. Whitt (1992) Ethics and Cultural Studies, in L. Grossberg *et al*, eds, *Cultural Studies*, Routledge, NY, 571-592.

Slethaug, B. (2000) *Beautiful Chaos: Chaos Theory and Metachaotics in Recent American Fiction*, SUNY Press, Albany.

Sobchack, V. (1990) A Theory of Everything: Meditations on Total Chaos, *Artforum* (November), 148-55.

Soule, M.E. (1995) The Social Siege of Nature, in M.E. Soule and G. Lease, eds, *Reinventing Nature: Responses to Postmodern Deconstruction*, Island, Washington, D.C. 45-56.

Spencer, H. (1876) *Principles of Sociology*, ed. by Andreski, 1969, from the original, Macmillan, London.

Spretnak, C. (1986) *The Spiritual Dimension of Green Politics*, Bear, Sante Fe.

Steiner, W. (1976) Language as Process: Serge Karchevskij's Semiotics of, in L. Matejka Sound, ed, *Sign and Meaning*, University of Michigan Press, Mich.

Steverson, B.K. (1994) Ecocentrism and Ecological Modeling, *Environmental Ethics*, 16, 71-88.

Sulis, W. and Combs, A., eds, (1996*) Nonlinear Dynamics in Human Behavior, World Scientific*, River Edge, NJ.

Sylvan, R. (1994) Illusion and Illogic in Evolution, *Revista Di Biologica*, 87, 191-221.

Sylvan, R. (1995) *Grand Philosophies and the Environmental Crisis*, unpublished paper.

Taylor, P.W. (1981), The Ethics of Respect for Nature, *Environmental Ethics*, 3, 197-218.

Tester, R. (1993) *The Life and Times of Postmodernity*, Routledge, London.

Thistle, D. (1981) Natural Physical Disturbance and Communities of Marine Salt Bottoms, *Marine Ecological Progam Series*, 6, 223-228.

Thompson, D., ed, (1996) *The Oxford Dictionary of Current English*, Oxford University Press, Oxford.

Tilby, A. (1993) *Science and the Soul: New Cosmology, the Self and God*, SPCK, London.

Tilman, D. (1982) *Resource Competition and Community Structure*, Princeton University Press, NJ.

Tilman, D. (1988) *Dynamics and Structure of Plant Communities*, Princeton University Press, NJ.

Tobey, R. (1981) *Saving the Prairies: The Life Cycle of the Founding School of American Ecology, 1895-1955*, University of California Press, Berkeley.

Trainer, F.E. (1985) *Abandon Affluence!* Zed Books.

Trojan, P. (1984) *Ecosystem Homeostasis*, Junk, Warsaw.

Turcotte, D. (1992) *Fractals and Chaos in Geology and Geophysics*, Cambridge University Press, NY.

Van Dyne, G.M. (1995) Ecosystems, Systems Ecology and Systems Ecologists, in B.C. Patten and S.I. Jorgensen, eds, *Complex Ecology*, (Paper repr. from 1966 version) Prentice-Hall, NJ, 1-21

Van Wyk, M. and Steeb, W. (1997) *Chaos in Electronics*, Kluwer Academic Publishers, Boston.

Vattimo, G. (1988) *The End of Modernity*, Polity Press, Cambridge.

Vaughn, K.I. (1987) *Austrian Economics in America: the Migration of a Tradition*, Cambridge University Press, Cambridge.

Volk, T. (1995) *Metapatterns Across Space, Time and Mind*, Columbia University Press, NY.

Volk, T. (1998) *Gaia's Body: Toward a Physiology of the Earth*, Copernicus, NY.

Von Frisch, K. (1967) *The Dance Language and Orientation of Bees*, Harvard University Press, Cambridge, MA.

Waldrop, M.M. (1993) *Complexity: The Emerging Science at the Edge of Order and Chaos*, Viking, London.

Waugh, P. (1992) Modernism, Postmodernism: Gender and Autonomy Theory, in P. Waugh, ed, *Postmodernism: A Reader*, Arnold, London, 189-204.

Waugh, P. (1992) Introduction, in P. Waugh, Ed, *Postmodernism: A Reader*, Arnold, London.

West, C. (1994) The New Cultural Politics of Difference, in S. Seidman, ed, *The Postmodern Turn*, Routledge, London, 65-81.

Westra, L. (1994) *An Environmental Proposal for Ethics*, Rowman and Littlefield, Baltimore.

Wheatley, M. (1992) *Leadership and the New Science: Learning About Organisation from an Orderly Universe*, Berret-Koehler Publishers, San Francisco.

White, E.C. (1991) Negentropy, Noise and Emancipatory Thought, in N.K. Hayles, ed, *Chaos and Order: Complex Dynamics in Literature and Science*, Chicago University Press, Chicago, 263-277.

White, P.S. (1979) Pattern, Process and Natural Disturbance in Vegetation, *Botanical Review*, 45, 229-299.

White, S.K. (1991) *Political Theory and Postmodernism*, Cambridge University Press, Cambridge.

White, D.R. (1997) *Postmodern Ecology: Communication, Evolution and Play*, SUNY Press, N.Y.

Whittaker, R.H. and G.M. Woodwell (1972) Evolution of Natural Communities, in J.A. Wiens, ed, *Ecosystem Structure and Function*, Oregon State University Press, Corvallis, OR, 137-59.

Wiens, J.A. (1984) On Understanding a Non-Equilibrium World: Myth and Reality in Community Patterns and Process, in D.R. Strong *et al*, eds, *Ecological Communities: Conceptual Issues and Evidence*, Princeton University Press, Princeton, 439-457.

Wiegert, R.G. (1988) Holism and Reductionism on Ecology: Hypothesis, Scale and System Models, *Oikos*, 53, 267-69

Wiener, N. (1961) *Cybernetics, or Control and Communication in the Animal and Machine*, MIT Press, MA. (First publ: 1949)

Wiessert, T.P. (1991) Representation and Bifurcation: Borge's Garden of Chaos Dynamics, in N.K. Hayles, ed, *Chaos and Order: Complex Dynamics in Literature and Order*, Chicago University Press, Chicago, 234-250.

Wilcox, D, (1996) What does chaos theory have to do with art? *Modern Drama*, 39, 4, 698-711.

Williams G.C. (1992) Gaia, Nature Worship and Biocentric Fallacies, *Quarterly Review of Biology*, 67, 470-483.

Worster, D. (1993) The Ecology of Order and Chaos, in S.J. Armstrong and R.C. Botzler, eds, *Environmental Ethics: Divergence and Convergence*, McGraw-Hill, NY, 39-48.

Worster, D (1994) *Nature's Economy: A History of Ecological Ideas*, [1st edn, 1977] Sierra Club, San Francisco.

Worster, D. (1995) Nature and the Disorder of History, in M.E. Soule and G. Lease, eds, *Reinventing Nature: Responses to Postmodern Deconstruction*, Island, Washington, D.C., 65-86.

Wright, R. (1996) Art and Science in Chaos: Contested Readings of Scientific Visualization, in R. Robertson *et al*, eds, *FutureNatural*, Routledge, London, 218-236.

Wu, J. and O.L. Loucks (1995) From Balance of Nature to Hierarchical Patch Dynamics: A Paradigm Shift in Ecology, *Quarterly Review of Biology*, 70, 439-466.

Yates, F.E. (1989) *Self-Organising Systems: The Emergence of Order*, Plenum Press, NY.

Young, R. (1974) The Historiographic and Ideological Contexts of the 19th Century Debates on Man's Place in Nature, in I. Teich and R. Young, eds, *Changing Perspectives in the History of Science: Essays in Honour of Joseph Needham*, Heinemann, London, 334-438.

Young, R. (1985) *Darwin's Metaphor*, Cambridge University Press.

Young, R. (1992) Science, Ideology and Donna Harraway, *Science as Culture*, 3, 2, 165-206.

Young, T. R. (1991). Chaos Theory and Symbolic Interaction Theory: Poetics for the Post Modern Sociologist. *Symbolic Interaction*, 14, 321-334

Zagorin, P. (1990) Historiography and Postmodernism: Reconsideration, *History and Theory*, 29, 263-75.

Zimmerman, M.E. (1994) *Contesting Earth's Future: Radical Ecology and PostModernity*, University of California Press, Berkeley.
Zimmerman, M.E. (1995) "The Threat of Ecofascism", *Social Theory and Practice*, 21 :207-238
Zohar, D (1990) *The Quantum Self: Human Nature and Consciousness Defined by the New Physics*, Morrow, NY.
Zohar, D. and I. Marshall (1994) *The Quantum Society: Mind, Physics and a New Social Vision*, Flamingo, London.

INDEX